I0541589

FROM COMBAT BOOTS

TO Blingy

HIGH HEELS

BY AMB. DR. LATASHA FENNELL (H.C.)

HAVANA BOOK GROUP LLC.
HAVANABOOKGROUP.COM

All rights reserved. No part of this publication may be reproduced, stored in a retrieval system, or transmitted in any form or by any means – electronic, mechanical, photocopying, recording or otherwise – without the written permission of the publisher.

HAVANA BOOK GROUP LLC
43537 RIDGE PARK DRIVE
TEMECULA, CA. 92590

COPYRIGHT 2025 All rights reserved.
ISBN: 9798992752502

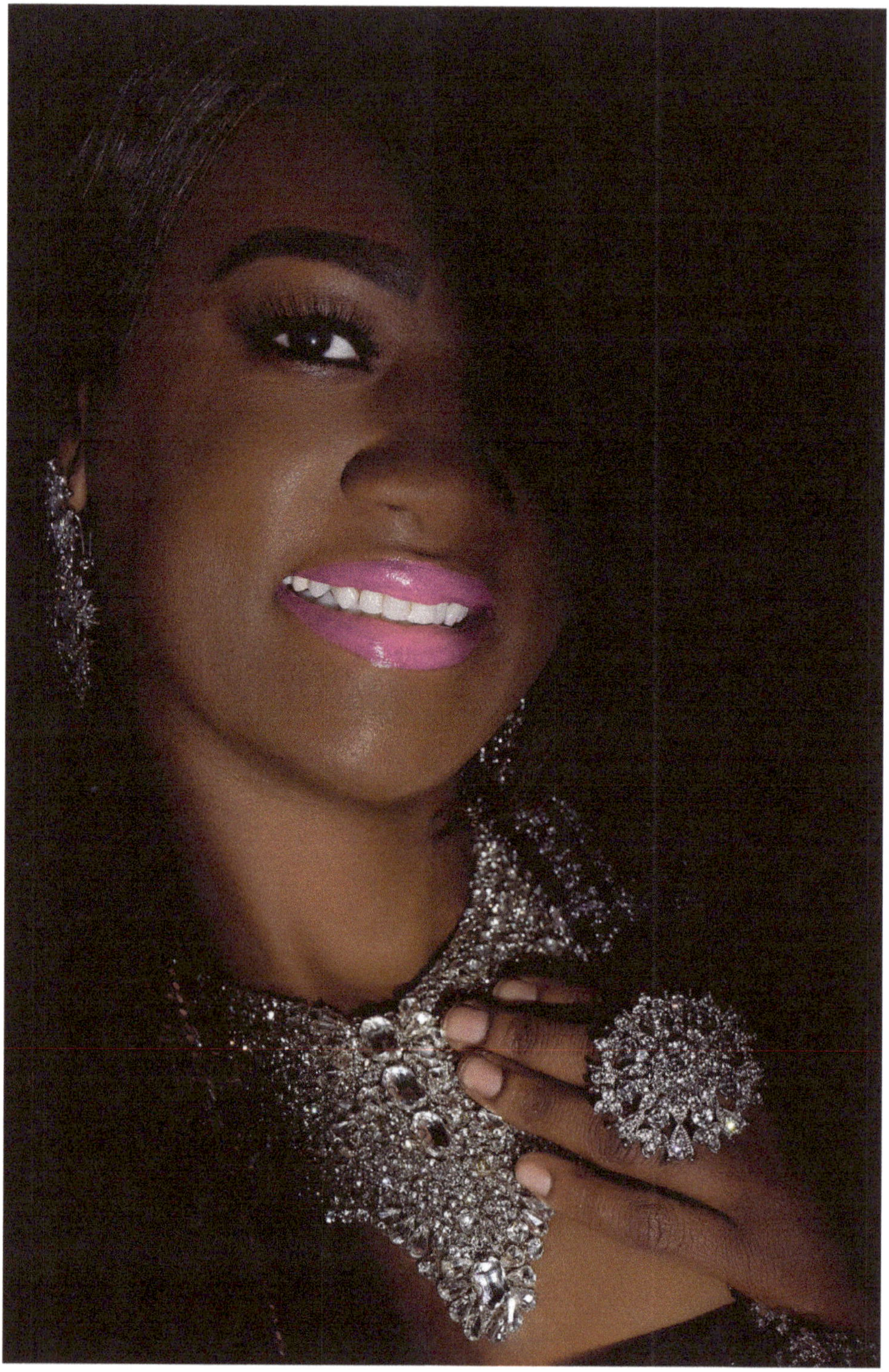

Dedication

This book is dedicated to all the strong, resilient women who have often been handed the short stick in life. To those who have made mistakes, carry flaws, and have fallen more than once—but never stayed down. You are the epitome of courage, grace, and determination.

You make the impossible look easy and the difficult look beautiful. You thrive unapologetically, carving out your place in this world with boldness and brilliance. This book is for you—the unstoppable forces making big moves, living boldly, and proving that every stumble in those blingy high heels is just another step toward greatness.

Here's to your strength, your perseverance, and your glow. Keep shining!

Amb. Dr. Latasha Fennell (h.c.)

Foreword

"From Combat Boots to Blingy High Heels," is a must-read book for the world!

The Journey of Ambassador Dr. Latasha Fennell whose life is a remarkable story of dedication, resilience, and self-reinvention. After serving for 25 years in the United States Navy as a high-ranking enlisted logistics specialist, she transitioned from prestigious military service to entrepreneurship, carving a niche for herself in the world of fashion. "From Combat Boots to Blingy High Heels" is a heartfelt memoir capturing her journey from the battlefields to the boardroom and from combat boots to stunning bling.

Foundation of Service

Latasha's journey began in her hometown, where she dreamed of making a difference. Enlisting in the Navy, she quickly learned the importance of discipline and teamwork as she adapted to military life. Her story follows her early years, detailing struggles and triumphs in training and the resulting camaraderie forged amid adversity.

Life in Logistics

As a logistics specialist, Dr. Fennell not only honed her skills in managing complex operations aboard ships, often in war zones, but also flourished as a leader. This book narrates her pivotal experiences, showcasing her contributions to crucial missions, the friendships that blossomed during challenging times, and the invaluable lessons learned about responsibility and resilience.

The Seeds of Entrepreneurship

Several years before her retirement, Latasha began envisioning her future outside of active duty. Understanding the need for a new direction, she took her love for fashion and started her Lady Bling business specializing in Navy-themed apparel, accessories, and stylish bling items. This book delves into her motivations behind launching her brand—how she sought to create not just beautiful products but a sense of belonging and pride among service members, veterans, and their families.

Bling on the Horizon

As she transitioned from military life to entrepreneurship, Dr. Fennell infused her business with passion and purpose, launching a range of bling clothing, hats, and jewelry that celebrates beauty.

Her book depicts her creative process, marketing strategies, and the hurdles she overcame in establishing her brand. With each successful sale, she shared a piece of her heart and a story of service, inspiring countless others along the way.

Embracing Motherhood

Alongside building her business, Dr Latasha navigated the challenges and joys of motherhood. Her book shares her role as a dedicated mother, balancing military service, entrepreneurship, and family life. Readers will be inspired by her tales of navigating this dual role; by showcasing the love and commitment she instilled in her family while empowering them to pursue their own dreams.

From Bling to Advocacy

Amb. Dr. Fennell's passion for service didn't end with her military career; she became a fierce advocate for veterans' rights and women's empowerment. This book details her initiatives and collaborations, using her platform as a successful entrepreneur and ambassador to drive meaningful change. Her commitment to advocating for her community is intertwined with her mission to inspire others to pursue their dreams, no matter the obstacles.

A Legacy of Empowerment

Amb. Dr Fennell reflects on her multifaceted journey, celebrating the path she paved from combat boots to high heels, embodying strength, elegance, and honor. She emphasizes the importance of embracing change, supporting one another, and leaving a lasting legacy. Amb. Dr. Fennell invites readers to see their struggles as steppingstones and to embrace their unique identities as they forge ahead.

"From Combat Boots to Blingy High Heels" is not just a memoir; it is an empowering narrative of transformation and resilience, encouraging readers to pursue their passions while celebrating their journeys. Amb. Dr. Latasha Fennell's story demonstrates that with determination and heart, anyone can navigate the complexities of life and emerge victorious, embodying both service and style.

Lady Amb Dr. Robbie Motter (h.c)
Founder /CEO:
Global Society for Female Entrepreneurs. A 501 (C)(3) nonprofit.
Contact: gsfeus.com

Preface

Every woman has a story. A journey of strength, resilience, and ambition. This book is mine—a reflection of my life, my struggles, and my triumphs as I transitioned from the structured discipline of combat boots to the creative freedom of blingy high heels.

I wrote From Combat Boots to Blingy High Heels to inspire and motivate women—small business owners, entrepreneurs, and everyday women who dream of something bigger. I want to show you that anything is possible. The key is to ask the right person, show up boldly, and never take "no" for an answer. If someone tells you "No," they're simply not the person who holds the key to your dreams.

My accomplishments didn't come easily. They were born from late nights, early mornings, and relentless determination. I've failed more times than I've succeeded, but I never stayed down. Instead, I surrounded myself with positive, powerful, and experienced women. I embraced failure as a steppingstone, using each setback to grow, learn, and build something better.

Many people see my achievements and assume I've had it easy or that I'm simply "lucky." The truth is, behind every success lies a story of struggle, sacrifice, and persistence. I've faced my share of challenges, and like many of you, I've had moments where I needed someone to lean on. Yet, I've always carried myself with grace, even during the hardest times, because I understood the importance of showing strength, even when I felt anything but.

This book is for the women who juggle countless roles and responsibilities—those navigating careers, business, motherhood, relationships, and life itself. It's for the women who fall but refuse to stay down. My hope is that my story inspires you to push forward, to dream bigger, and to take bold steps toward your goals.

I share my journey, not as a blueprint, but as a testament to what's possible when you believe in yourself. Learn from my failures, my successes, and my unwavering belief that you can achieve greatness. Set goals, create timelines, and never be afraid to dream so big that it scares you—and everyone around you.

Remember, every failure is a lesson, and every struggle is a chapter in your story. This is my story. Let it be a spark that ignites your own journey toward a life filled with purpose, passion, and success.

With love and blingy inspiration,

Amb. Dr. Latasha Fennell (h.c)

Introduction

From Combat Boots to Blingy High Heels is more than just a book—it's a journey of resilience, transformation, and unapologetic self-expression. As a proud 25-year veteran of the United States Navy, I've experienced the challenges, triumphs, and transitions that come with serving my country and stepping into a new chapter of life. This book is my personal story of trading in my combat boots for blingy high heels, blending the discipline and strength of military life with the creativity and sparkle of entrepreneurship.

Throughout these pages, I'll share the pivotal moments that shaped me—lessons learned on the battlefield, the joys and hurdles of retirement, and the passion that led me to create Boss Lady Bling Blingy Boutique. But this book isn't just about my story. It's for every woman who has ever faced a transition and wondered, What's next? It's for the dreamers, the doers, and the trailblazers who are ready to embrace their sparkle while staying true to their strength.

I've dedicated my post-military career to empowering others, whether it's through designing custom, one-of-a-kind pieces, mentoring veterans as they navigate their new lives, or sharing the spotlight with those who need it most. This book is an extension of that mission—a love letter to transformation and a guide to finding your confidence in every step, whether you're marching in boots or strutting in stilettos.

Thank you for walking this journey with me. Together, let's celebrate the beauty of reinvention and the courage it takes to shine.

With love and bling,

Amb. Dr. Latasha Fennell (h.c.)

Boss Lady Bling Blingy

Table of Contents

Foreword

Preface

Introduction

Chapter 1: The Girl Behind the Boots

I am just a small-town country girl with big-city dreams. Born in Memphis, Tennessee, I grew up in a rural community called Anthony Quarters, just outside of West Memphis, Arkansas. It was a small but big place, with two main rock roads leading into town. Anthony Quarters was the kind of place where everyone knew everyone, and dreams often got swallowed by the weight of reality. But even as a young girl, I knew I was destined for something greater.

Life was not always kind. It threw me curveballs, some that knocked me down hard. But I had a fire in my soul and faith in God to guide me. From an early age, I learned how to set goals, fight for them, and never give up, even when the odds were not in my favor. There were times when I felt stuck, broken, and unsure of my next step. Yet, through every setback, I chose resilience over resignation, knowing that each trial was shaping the woman I was meant to become.

Growing up in Anthony Quarters required grit. My parents divorced when I was about twelve years old, a heart-wrenching shift that left me grappling with a sense of loss. Both of my parents were hardworking and devoted, instilling in me and my siblings the importance of diligence, education, and ambition. They believed we could achieve remarkable things, even though they hadn't attended college themselves. In our little community, higher education was more a dream than a standard path, and most people worked modest jobs to make ends meet. For me, finishing school and finding a 9-to-5 job seemed like the natural roadmap. College wasn't something I thought I could even do, let alone pursue.

Anthony Quarters may have been small, but it had a character all its own. We had one fire truck, a funeral home, and a small store run by Mr. Bruce, who seemed to have everything we needed. If you needed milk, bread, or a friendly ear, you went to Mr. Bruce's store. It was more than just a shop; it was the heart of the community.

Then there was the Juke Joint—a lively club where people from all around came to dance and let loose. On weekends, the music would echo through the night, and I would sneak close to watch. The women were mesmerizing in their shiny outfits, fancy hairstyles, and glittering jewelry. They seemed so confident, so full of life, and they carried themselves like queens. Though I was supposed to be inside when the streetlights came on—and we only had about five in the whole town—I would steal a few moments to sit outside, sway to the music, and dream of a life as glamorous as theirs.

Summers in Anthony Quarters were filled with hard work and valuable lessons. I chopped cotton in the fields, helped tend to the garden, and pitched in on the family's small farm. But some of my favorite memories were with my grandmother, who was the cornerstone of our family. She was the kind of woman who could do anything and do it well. She taught me how to cook, sew, and even sparked my first interest in entrepreneurship. My grandmother was the "candy lady" of the community, and I often helped her sell homemade treats, Sunday dinners, and her famous hot tamales. People would drive down from other towns just to get a taste of her cooking. She worked tirelessly, but she always made it look effortless.

Those days spent under the watchful glow of the streetlights or in the bustling kitchen with my grandmother, planted the seeds of who I would become. Anthony Quarters wasn't just a small town with two dusty rock roads—it was where I learned about resilience, hard work, and the power of dreams.

The girl behind the boots may have started her journey in a tiny town where dreams seemed out of reach, but I always knew there was more waiting for me beyond those rock roads. Each challenge I faced, each lesson I learned, and each dream I dared to dream was shaping the woman I was destined to become.

A Life-Changing Encounter

The summer I turned eleven was like any other in the countryside—peaceful, predictable, and full of simple routines. One of my weekly chores was bringing in the laundry from the clothesline, where it hung to dry under the warm sun. I'd done it countless times, pulling in the fresh, crisp clothes, folding them, and putting them away. But that day, something unexpected turned an ordinary task into a life-altering moment.

As I brought in the clothes, I grabbed a shirt and decided to slip it on. It seemed harmless enough—a habit I'd done hundreds of times without a second thought. But the moment the fabric touched my skin, I felt a sharp, burning pain on my left arm. Instinctively, I panicked, swatting at the source. As I struggled to get the shirt off, the spider hidden in its folds became enraged, biting me a second time before scurrying away.

The pain was immediate and intense, spreading like wildfire through my arm. By the time my parents saw the swelling and discoloration, it was clear something was very wrong. They rushed me to the hospital, where the doctors confirmed the worst—it was a brown recluse spider bite. And not just one, but two bites meant the venom was working faster, spreading through my arm with devastating speed.

The doctors' faces were grim as they explained what was happening. The venom had caused necrosis, killing the tissue in my arm. Their recommendation was amputation. Even worse, they said that even if the arm could be saved, I would likely never regain full use of it.

I was devastated. At eleven years old, I couldn't imagine a life where I couldn't use my left arm the same way again. I refused to believe that this was my future. Sitting in that cold hospital room, I begged the doctors and my parents to find another way. "There has to be something else we can do," I pleaded, tears streaming down my face. I wasn't ready to let go—not of my arm, not of my dreams, and not of the life I had imagined for myself.

What followed were months of grueling surgeries, treatments, and recovery. After ten surgeries later, my arm was still attached, though deeply scarred. Each procedure felt like a small war—one where I had to summon courage, I didn't know I had. The physical pain was unbearable, but the emotional weight of wondering if I'd ever be the same was even harder to bear.

The doctors' words echoed in my mind: "You'll never use your left arm the same way again." But something in me refused to accept that as my truth. With determination and the unwavering support of my family, I worked through the pain and the fear. Slowly, I began to heal, defying the odds one small step at a time.

Finding My Direction

Returning to school after months in the hospital was a victory I'll never forget. I was behind in my studies and self-conscious about my scars, but I was determined to catch up. My classmates stared at the marks on my arm, but I didn't let their curiosity bother me. Instead, I wore my scars proudly, as symbols of my resilience and strength.

That spider bite changed my life in ways I couldn't have imagined. It taught me that strength isn't just about physical endurance—it's about refusing to give up, even when the odds are stacked against you. Despite what the doctors had said, I found ways to use my arm again, even if it wasn't exactly as it had been before.

Looking back, I realize that what could have been the end of so many dreams became a beginning instead. The bite left scars on my arm, but it also left a lasting mark on my spirit, reminding me every day that no obstacle is too great to overcome. That experience shaped the person I would become, showing me that faith, determination, and an unyielding spirit can overcome even the toughest challenges.

The Navy Calls

After high school, I stood at a crossroads. College felt out of reach, both financially and emotionally. But the United States Navy? That was something different. It offered adventure, discipline, and a chance to see the world beyond Arkansas. Joining the Navy wasn't the future my parents had envisioned, but deep down, I knew it was the path I needed to take.

Enlisting wasn't easy. My past came back to haunt me during the medical screening. The doctors took one look at my medical history and hesitated. My arm—the same one I had fought so hard to save—became the barrier between me and my dreams again. For a moment, it felt like the world was telling me I wasn't good enough.

But I refused to accept that. I fought, just as I had years ago in the hospital. I underwent evaluations, tests, and more hoops than I thought possible. And when the moment came, I passed.

Enlisting in the Navy was one of the proudest moments of my life. It wasn't just about wearing the uniform or traveling the world—it was about proving to myself and everyone else that no battle, physical or emotional, could keep me from rising above.

The Girl Behind the Boots

As I laced up my combat boots for the first time, I reflected on the journey that had brought me there. A small-town girl, scarred but unbroken, stepping into a future that held promise and purpose. Those boots weren't just military-issued footwear—they symbolized every hurdle I had leaped over, every tear I had shed, and every victory I had earned.

I joined the Navy to discover who I was and what I was capable of. But in reality, I was building the foundation for a story that would one day inspire others. Because the girl behind the boots? She's a fighter, a dreamer, and a woman who refuses to let life's challenges define her.

EIGHTH GRADE

-M-M Good!

re Thanksgiving the home
ics department started a
er. The fundraiser was
cookies. Flavors available
ocolate chip, butter sugar
ite chocolate macadamia
e cookies were furnished
. Spunkmeyer. Otis
neyer also furnished two
n which to bake the
. The cookies were sold
h the first week of
er. Mrs. Weaver said,
oney made will be used
e economics projects, the
rial Arts Christmas Party,
uch more.

economics student Melinda
imbreea is taking out a new batch
es. She was one of the students
ped bake and sell cookies during
a fever.

Jessica Meyers
Sharvell Minter
Rachel Moore
Brooke Morgan
James Mosley
Latasha Motley
Daniel Murphy

Misty Nichols
Jason Pedigo
Chara Piney
Kristen Person
Josh Pixley
Jack Poit
Chrissa Powell

Clereth Puschett
Jamarius Puschett
Keon Puwito
Lazarus Price
Melody Reynolds
Shayne Rhyne
Sharene Rhyne

Cindy Ridge
Justin Rikard
Drew Roberts
Chad Rogers
Charlotte Rolfe
Vickie Sanders
Ronnie Schroeder

Bruce Scott
Matthew Scott
Justin Scrape
Allison Sheffield
Derick Simmons
Franklin Simpson
Carmen Skillern

Elliott Sloan
Brianne Smith
Sammy Smith
Stephanie Smith
Jim Snyder
Maria Spitz
Beth Stails

Chris Stanton
Michael Starkey
Stephanie Stephens
Josh Stewart
Jonathan Stevenson
Antwain Stinnett
Derrick Strane

James Mosley 7
Brandon Moss 8
Elizabeth Moss 8
Latosha Motley 7
J Manolewhite 9
Jyied Myers 8
Muzi Myrick 7

Hunter Naylor 8
Adam Neal 9
Judy Neal 8
Brad Nelson 9
Donald Nelson 8
Jason Newsom 9
Misty Nichols 7

Marla Nichols 8
Matthew Odom 9
Blake Owen 9
Crystal Parker 8
Brook Patterson 9
Darrin Paterson 8
Melinda Patterson 9

Allana Paterson 8
James Pedigo 7
Shane Pedigo 8
Brandon Perrine 8
Chara Perry 7
Barrett Person 9
Kristen Person 7

Chapter 2: Basic Training: Breaking Down and Building Up

"Welcome to Great Lakes, Illinois," the bus driver announced as we rolled up to Recruit Training Command. Despite the calmness in his voice, my heart was pounding. This was it—the beginning of my journey in the United States Navy. As I stared out the window into the chilly Chicago night, I tried to imagine what the next few weeks would bring. The bus was silent except for the hum of the engine, but tension hung heavy in the air. None of us knew exactly what to expect, and our collective nervousness seemed almost tangible.

The first few hours were a blur of barking voices, hurried instructions, and the realization that my life had officially changed. I was no longer just me—I was now a recruit, one of many. My hair was quickly tied back as neatly as possible, my clothes swapped for the standard-issue navy blue sweatpants and hoodie, and my civilian identity tucked away like contraband. Even the smallest comfort, like my personal belongings, was stripped away, leaving me with nothing but the essentials: a standard-issue backpack, a pair of black boots, and my determination to make it through.

The Fear of Water

One of the first hurdles I encountered was the swimming test. For most recruits, it was an expected but manageable part of training. For me, it was a nightmare. I had grown up fearing water. In my mind, swimming equated to drowning, and no amount of reasoning seemed to make that fear go away. On the day of the test, I stood at the edge of the Olympic-sized pool, my knees trembling. "Jump in!" the instructor shouted. My heart raced as I stared into the water's depths.

When I finally mustered the courage to leap, my body sank like a stone. Panic took over, and I flailed wildly, gasping for air as I struggled to stay afloat. But the instructors weren't there to coddle us. "You're not getting out of here until you learn to swim," one of them bellowed. It was harsh, but it was the reality I needed to face. Over the next several days, I spent every spare moment practicing in the pool. My fellow recruits cheered me on, offering tips and encouragement. Slowly but surely, I learned to tread water, then float, and finally swim. By the end of training, I not only passed the test but overcame one of my deepest fears—a victory that remains one of my proudest moments.

The Art of Breaking You Down

Boot camp wasn't just about physical challenges—it was about stripping us down to our rawest selves. Everything about the experience was designed to test and break us. The early mornings started with the blaring sound of a whistle, jolting us awake at 4:30 a.m. From there, it was nonstop: rigorous physical training, marching drills, classroom lessons, and more physical training.

But it wasn't the endless push-ups or mile-long runs that truly broke me down. It was the mental and emotional strain. Sleep deprivation, constant correction, and the pressure to succeed weighed heavily on all of us. I remember one night lying in my rack, staring at the ceiling, and wondering if I had made the biggest mistake of my life. "What am I doing here?" I thought.

The hardest part was the isolation. Letters from home were few and far between, and phone calls were even rarer. I missed the comfort of my mother's voice, the smell of my favorite home-cooked meals, and the familiar sights of my hometown. I had traded all of that for a life of uncertainty, and in those quiet moments of doubt, the enormity of my decision sank in.

Building You Back Up

But just as boot camp was designed to break us down, it was also designed to build us back up. Slowly, I began to realize that every hardship had a purpose. The relentless schedule taught me discipline. The grueling physical demands made me stronger than I ever thought possible. The constant correction forced me to let go of my ego and embrace teamwork.

One pivotal moment came during an endurance run. My body felt like it was on the verge of collapse, my legs screaming in protest with every step. I wanted to quit so badly that tears welled up in my eyes. But then I heard the voices of my fellow recruits. "You got this!" one shouted. "Come on, we're almost there!" another chimed in. Their encouragement lit a fire in me, and I pushed through the pain to finish the run.

That was the day I learned the true meaning of resilience. It wasn't about being the fastest or the strongest—it was about refusing to give up, no matter how hard it got. Boot camp taught me that resilience is a muscle, one that grows stronger each time you choose to persevere.

A New Kind of Family

One of the most unexpected rewards of boot camp was the sense of camaraderie that developed among the recruits. We came from all walks of life—different races, religions, and backgrounds—but in that shared struggle, we found common ground. There's something about enduring hardship together that forges bonds stronger than steel.

I'll never forget the night our division came together for the first time. We were practicing a drill routine, and nothing seemed to be going right. Tempers flared, and frustration was high. But instead of letting it divide us; we decided to talk it out. For the first time, we shared our fears, our struggles, and our hopes for the future. That night, something shifted. We stopped seeing each other as competitors and started seeing each other as teammates.

From that point on, we became a family. We cheered each other on during training, shared our snacks during rare moments of downtime, and supported each other when the pressure became too much. That bond carried us through the toughest days and reminded us that we weren't in this alone.

Finding Myself

Perhaps the most profound lesson I learned in boot camp was about myself. Before joining the Navy, I had carried a lot of baggage—insecurities about my appearance, unresolved pain from my childhood, and a lingering fear of failure. I had always been good at hiding those feelings, but boot camp didn't leave room for pretense.

When you're pushed to your limits, your true self comes to the surface. I couldn't hide my fears, my weaknesses, or my emotions. But instead of being judged, I found acceptance. My fellow recruits saw me for who I was and supported me anyway. That acceptance gave me the courage to confront my insecurities and start working through them.

By the time I graduated, I felt like a different person. I was stronger, more confident, and more self-assured. I had faced my fears, embraced my strengths, and discovered a resilience I didn't know I had.

The Freedom to Grow

Graduating from boot camp wasn't just a milestone—it was a turning point. It marked the beginning of a journey that would take me around the world, introduce me to incredible people, and open my eyes to new possibilities. For the first time, I felt like I was truly in control of my destiny.

The Navy gave me more than just a career—it gave me the freedom to grow. It gave me the tools to push past my limits, the courage to chase my dreams, and the strength to overcome any obstacle.

As I stood at attention during my graduation ceremony, listening to the cheers of my fellow sailors and our families, I felt an overwhelming sense of pride. I wasn't just a small-town girl with big dreams anymore. I was a United States Navy sailor, ready to take on the world.

Chapter 3: My First Deployment - A New World

In 1999, my world shifted forever. I had just completed boot camp, a grueling, transformative experience that challenged me physically, mentally, and emotionally. Afterward, I moved on to A School in Meridian, Mississippi, where I was trained as a Storekeeper (SK). The idea of managing budgets, purchasing supplies, and handling inventories felt like a perfect fit. Numbers had always made sense to me, and I enjoyed the satisfaction of organization and efficiency. In A School, I envisioned my future in the Navy as a seamless blend of practicality and purpose—a career where I could thrive. But as I would soon discover, expectations and reality often diverge in the most surprising ways.

When I received my orders for my first duty station, my excitement was palpable. I was assigned to the United States Navy's Mobile Construction Battalion Three (NMCB-3), better known as the Seabees. Based in Port Hueneme, California, the battalion's reputation preceded it: rugged, hardworking, and unyielding. The Seabees were builders, innovators, and problem-solvers. Their motto, "We build, we fight," captured the essence of their mission. I was stepping into a male-dominated environment filled with individuals who had earned their place through grit and determination.

Port Hueneme was nothing like what I had imagined my military life would be. The base was a stark contrast to the structured environment of A School. It was tough and gritty, a place where field exercises were as common as

routine paperwork. I quickly realized that being a part of the Seabees meant more than just handling finances—it meant understanding the core of what it took to keep a battalion running. The job was serious business, and there was no room for error.

Facing Challenges and Doubts

Transitioning into active duty was like learning to swim by being thrown into the deep end. I went from the structured training of A School to the chaotic reality of managing the battalion's OPTAR (Operating Target) budget. The numbers weren't just abstract figures; they represented the lifeline of our operations. Every purchase—from tools and uniforms to fuel and equipment—had to be meticulously accounted for. The stakes were high, and the pressure was immense.

There were moments of doubt, times when I felt overwhelmed and out of place. As one of the few women in the battalion, I had to prove myself not just as a competent Storekeeper but as someone who could keep up with the demanding pace of Seabee life. The thought of failing terrified me, but it also fueled my determination. I reminded myself why I joined the Navy in the first place—to challenge myself, to grow, and to serve a purpose larger than myself.

Then came the announcement that would change everything: NMCB-3 was deploying to Guam. I was barely settled into my role, still trying to find my footing, and now I was preparing to go overseas for seven months. The news filled me with a mix of excitement and fear. This would be my first deployment, my first time traveling to a foreign land, and my first major test as an active-duty sailor. I was stepping into the unknown, and the weight of that realization was both thrilling and terrifying.

The Journey to Guam

The journey to Guam was an adventure in itself. Traveling aboard a C-130 aircraft was unlike anything I had ever experienced. The noise was deafening, the seats were cramped, and the flight was far from comfortable, but the anticipation of what lay ahead kept my spirits high. As we descended toward the island, I caught my first glimpse of Guam's lush greenery and sparkling waters. It was breathtaking, a stark contrast to the military base I had left behind.

But as beautiful as Guam was, it was no vacation destination for us. The Seabees had a mission, and that mission required focus, discipline, and hard work. The tropical heat was relentless, the humidity clinging to us like a second skin. Every day was a test of endurance, both physically and mentally. I quickly learned that deployment life was unpredictable, and adaptability was key to survival.

Finding Strength Amid Challenges

Guam brought its own set of challenges. The work was demanding, often pushing me beyond what I thought I could handle. There were days when I felt homesick, longing for the familiarity of family and friends. But there were also moments of triumph, times when I felt a newfound strength emerging within me. Each obstacle I faced became an opportunity to grow.

One of the most significant challenges was earning the respect of my peers in such a male-dominated environment. The Seabees prided themselves on their toughness and resilience, and as a young woman, I often felt the need to prove myself twice as hard. But as I immersed myself in my role, I started to see the impact of my work. Managing the battalion's financial resources was no small task and knowing that my efforts contributed to the success of our mission gave me a sense of pride and purpose.

The camaraderie I found with my fellow Seabees was another source of strength. Despite our differences, we were united by a common goal. We worked together, supported one another, and shared both the struggles and the victories of deployment life. They became like family to me, and their support helped me navigate the toughest days.

Discovering Guam's Beauty and Culture

Amid the challenges, there were moments of joy and discovery. Guam's natural beauty was awe-inspiring, from its pristine beaches to its vibrant sunsets. On rare days off, I had the chance to explore the island, experiencing its rich culture and warm hospitality. I learned about Chamorro traditions, tasted local delicacies, and marveled at the resilience of a people who had endured so much throughout their history.

These moments of connection and cultural exchange reminded me of why I had joined the Navy in the first place. It wasn't just about serving my country—it was about experiencing the world, learning from others, and broadening my horizons.

A Turning Point

Looking back, my first deployment to Guam was a turning point in my life. It was more than just an introduction to the realities of military service; it was a journey of self-discovery. I learned to face my fears head-on, to adapt to new and challenging environments, and to push myself beyond my perceived limits.

Guam taught me the value of resilience and the importance of teamwork. It showed me that strength comes not from avoiding challenges but from embracing them. And it proved to me that I was capable of far more than I had ever imagined.

Lessons for Life

The lessons I learned during those seven months in Guam stayed with me throughout my military career and beyond. They shaped me into a stronger, more confident individual, someone who could tackle challenges with determination and grace. My first deployment was the beginning of a journey that would take me to places I never dreamed of, both physically and emotionally.

As I reflect on that time, I am filled with gratitude for the experiences that shaped me and the people who stood by my side. Guam will always hold a special place in my heart—not just as my first deployment but as the place where I discovered the strength within myself to face whatever life might throw my way.

That deployment was more than just a chapter in my Navy career; it was the foundation of a life defined by resilience, purpose, and a deep appreciation for the journey.

Chapter 4: Making Rank

Achieving a new rank in the Navy is more than just a promotion; it's a testament to hard work, perseverance, and the ability to adapt under pressure. When I was announced as SK3—Storekeeper Petty Officer Third Class—what many outside the Navy know it as Petty Officer, my excitement was indescribable. It wasn't just the new title; it was the acknowledgment that my dedication and efforts were being recognized. But with that recognition came a new reality— being promoted meant stepping up as a leader, embracing responsibility, and navigating challenges that would shape me both personally and professionally.

The Privileges of Promotion

The saying "Rank has its privileges" is something every sailor quickly learns. Moving up in rank brought benefits that were both practical and symbolic. For instance, gone were the days of being crammed into a small living space with three or four roommates during deployment. As an SK3, I was granted a two-person room on deployment. It may not seem like much to some, but having fewer people in your space was a massive upgrade—less noise, more privacy, and room to breathe.

Even more exciting was the opportunity to live alone in homeport. With a simple chit—essentially a formal request— you could apply for Basic Allowance for Housing (BAH) and move off base to live in town. While I wasn't financially ready to take that step just yet, the prospect of having my own room was thrilling enough. After years of sharing, the thought of a space to call entirely my own felt like a luxury.

But rank wasn't just about personal perks. With it came higher expectations, greater responsibility, and the opportunity to develop leadership skills that would prepare me for future challenges.

The Weight of Responsibility

Becoming an SK3 wasn't just a career milestone—it marked the beginning of my journey as a leader. As a junior petty officer, I was no longer just following orders; I was now giving them, mentoring younger sailors, and ensuring tasks were executed efficiently. I had to set the example—arriving early, staying late, and always maintaining professionalism.

Leading under pressure was a lesson I had to learn quickly. Whether it was managing supplies during a high-stakes operation or ensuring my team met deadlines despite obstacles, I realized leadership wasn't just about telling others what to do. It was about listening, problem-solving, and staying calm when things didn't go as planned.

One of my first challenges came when I was tasked with inventory management for a critical mission. Missing even one item could derail operations. I had to ensure everything was accounted for, even if it meant staying up late to triple-check the paperwork or helping my team organize supplies. There were moments of frustration, but I learned that leadership often requires putting others' needs before your own comfort.

Earning Respect

Leadership isn't given; it's earned. As an SK3, I had to prove myself to both my superiors and my peers. I discovered that respect came not just from completing tasks but from how I carried myself. Taking the time to mentor younger sailors, offering support when they struggled, and showing gratitude for their hard work created a bond of mutual respect.

One moment that stands out was when a junior sailor came to me with a personal issue that was affecting their performance. Instead of brushing them off, I took the time to listen and help them find a solution. That conversation not only helped them get back on track but also strengthened their trust in me as a leader.

The Bigger Picture

Earning my promotion to SK3 was a reminder of why I joined the Navy in the first place. It wasn't just about personal success—it was about contributing to something greater than myself. The Navy's mission, the camaraderie of my shipmates, and the pride of wearing my uniform drove me to keep pushing forward.

Making rank taught me the importance of resilience, adaptability, and the power of leading with integrity. It wasn't always easy, and there were days when the pressure felt overwhelming. But every challenge was an opportunity to grow, and every success reinforced my belief in my own potential.

As I moved into this new chapter of my career, I realized that leadership isn't about being perfect; it's about being present, being accountable, and being willing to learn. Making rank was just the beginning, and I was ready to embrace the journey ahead.

Chapter 5: Personal Sacrifices of Service

It was in my early years of service in the United States Navy when I met the man who would become the love of my life. At the time, I was stationed with NMCB-3, part of the Seabee battalion. There, amidst the demands of our duty, I crossed paths with a young man whose warm, brown eyes were unforgettable. Though we came from two different worlds— he, a proud Jamaican with a rich cultural heritage, and I, a determined young woman chasing my dreams in the military— our differences only strengthened our bond.

We began dating in Port Hueneme, California, a chapter in my life that felt like a whirlwind of excitement and new beginnings. He quickly became my Jamaican King, the one I could rely on no matter the challenges we faced. Together, we built memories filled with laughter, adventure, and a shared determination to conquer the hurdles life threw at us.

Our journey took us across the world, and when we were deployed together to Rota, Spain, it felt like a dream come true. The experience of traveling and discovering the beauty of Spain with him by my side was magical. Rota became the backdrop to some of the happiest moments of my life, and it was there, amidst the vibrancy of Spanish culture, that we created something even more beautiful—a life.

The news of my pregnancy brought with it a surge of emotions—excitement, joy, and a deep sense of fear. Military life, particularly in the same unit, often frowned upon relationships and pregnancies. There I was, pregnant overseas, juggling my duties as a sailor while navigating the uncharted waters of impending motherhood.

And then the world changed.

September 11, 2001, will forever be etched in my memory as one of the most tumultuous times in my life. The tragedy shook not only our nation but also my small corner of the world. Rota, once a haven of joy and exploration, became a place of uncertainty and heightened tension. At eight months pregnant, I found myself trapped in a foreign land, far from home, wrestling with choices that would alter the course of my life.

The Navy demanded unwavering commitment, but I was now a mother-to-be with a growing desire to nurture and protect the life within me. The decision was one of the hardest I've ever had to make: Should I continue my career in the military, or should I leave to dedicate myself fully to raising my son?

In the end, I chose to do both.

The journey was anything but easy. Serving in the military meant that there would be times I'd have to leave my son and husband behind to fulfill my duties. It meant missing milestones, birthdays, and moments I could never get back. But it also meant providing a stable future for my family, serving a cause greater than myself, and demonstrating to my son what resilience and commitment look like.

Through it all, I leaned on prayer, the unwavering support of my husband, and the love of my family. It was their belief in me that gave me the strength to press forward.

The Struggles of Balancing Two Worlds

Balancing military life with personal relationships is no small feat. The Navy demands discipline, time, and often, sacrifice. Deployments pulled me away for months at a time, leaving my husband to shoulder the responsibilities of parenting alone. He never complained, but I could see the toll it took on him.

My heart broke every time I boarded a plane or ship, knowing I was leaving behind a piece of myself. I remember vividly the tears in my son's eyes when he was old enough to understand I was leaving, but not quite old enough to understand why. He would cling to me, his small hands gripping my uniform, begging me to stay. Those moments shattered me in ways I can't fully describe.

At times, guilt became a constant companion. I questioned if I was doing enough for my family, if my sacrifices were worth the pain they endured in my absence. I wrestled with the fear that my son might grow up resenting my service, feeling as though I chose my career over him.

The Toll on My Marriage

My marriage faced its own set of challenges. My husband, though endlessly supportive, had his moments of frustration and loneliness. The long months apart, the missed anniversaries, and the constant readjustments whenever I returned home tested the strength of our bond.

Communication became our lifeline. We learned to cherish every phone call, every letter, and every fleeting moment of connection. We became experts at navigating the complexities of a long-distance relationship, but it was far from easy.

Still, my husband's unwavering belief in me kept me grounded. He reminded me of my strength, my purpose, and the impact I was making not just on our family but on the world. He sacrificed his own dreams and desires to support mine, and for that, I am eternally grateful.

Finding Strength in the Struggle

Despite the challenges, I found strength in my faith and my purpose. I knew that my service was not just for me but for something greater. It was for my son, to show him that his mother was a fighter, a leader, and someone who never gave up. It was for my husband, to prove that our love could withstand anything. And it was for the countless others who served alongside me, who understood the sacrifices we made to protect the freedoms we hold dear.

Over time, I learned to find balance. I learned to be present in the moments I had with my family, to make the most of every opportunity to show them love and appreciation. I learned to forgive myself for the times I couldn't be there, trusting that the foundation I was building for them would make up for my absence.

The Legacy of Sacrifice

Looking back, I see my sacrifices not as burdens but as building blocks of my legacy. My son grew up knowing the value of hard work, resilience, and dedication. My marriage, though tested, became stronger and more rooted in mutual respect and love.

Every challenge I faced, every tear shed, and every moment of doubt was worth it. They shaped me into the person I am today—a woman who embodies strength, perseverance, and unwavering love for her family.

Conclusion

Balancing military life with personal relationships is a constant dance of sacrifice and love. It is not for the faint of heart, but it is a journey filled with profound lessons and unbreakable bonds.

For me, the toll it took on my heart was heavy, but the rewards were immeasurable. I am proud of the choices I made, the sacrifices I endured, and the legacy I built for my family. My story is a testament to the resilience of the human spirit and the power of love to endure even the most challenging circumstances.

As I continue to reflect on my journey, I am reminded that the sacrifices of service are not just mine—they are shared by every member of my family. Together, we have faced the trials of military life and emerged stronger, closer, and more grateful for the blessings we have.

Chapter 6: Being a Mom in Uniform:

I still remember the day I found out I was going to be a mom. It was a mix of excitement, fear, and disbelief. At that moment, I didn't fully grasp how much my life was about to change. As a proud Petty Officer Second Class in the United States Navy, I was used to structure, discipline, and long hours. But when I became a mother, my life's mission shifted. Suddenly, I wasn't just serving my country; I was serving a tiny human who depended on me for everything.

Balancing motherhood and military service was one of the hardest challenges I've ever faced. It was messy, exhausting, and overwhelming. But through the chaos, I discovered a strength I never knew I had.

The New Reality

In the Navy, every day is a mission. There are inspections to pass, tasks to complete, and standards to uphold. But now, I had another mission: being a mother. My days started at 4:30 a.m., with barely enough time to get ready before waking my baby. Diaper changes, feedings, and daycare drop-offs all had to be done before I reported for duty.

At work, I was a leader, responsible for mentoring sailors and completing critical tasks. I had to stay sharp, competitive, and professional. But my mind was often elsewhere, worrying about my baby. Did he cry when I left him at daycare? Was he okay with the sitter?

By the time I got home, exhaustion had set in. But there was no time to rest. Dinner needed to be made, laundry piled up, and bottles had to be washed. Then came the bedtime routine—a nightly battle of rocking, singing, and praying he would fall asleep quickly so I could get a few hours of rest before doing it all over again.

The Weight of Doing It Alone

One of the hardest parts of this journey was the loneliness. My family lived thousands of miles away, and I didn't have the luxury of calling on grandparents or siblings for help. My husband, also in the military, was often deployed, leaving me to figure things out on my own.

There were nights when I would collapse on the couch, tears streaming down my face from sheer exhaustion. I wanted to call my mom, my best friend, or anyone who could listen and tell me it would be okay. But I didn't want to burden them with my struggles.

The weight of responsibility was overwhelming. I was a mother, a sailor, and a wife. Each role demanded so much of me, and I felt like I was barely holding it together.

The Tough Decisions

Motherhood forced me to make difficult decisions. While my friends were out enjoying their twenties, going to clubs and traveling, I was changing diapers and paying bills. I felt like I was missing out on so much. But at the same time, I wouldn't trade my baby's giggles or tiny hugs for anything in the world.

There were times when I had to say no, to opportunities at work because I couldn't afford to stay late or volunteer for extra duties. I worried that my peers were advancing faster than I was, and I wondered if I was jeopardizing my career.

But I also knew that my priorities had changed. My son came first. He was my reason for working so hard and my motivation to keep going, even on the hardest days.

Faith as My Anchor

Through it all, my faith kept me grounded. There were nights when I felt like giving up, but I would sit down, close my eyes, and pray. I asked God for strength when I felt weak, for patience when I was overwhelmed, and for guidance when I didn't know what to do.

One night, after an especially rough day, I opened my Bible and came across a verse that stayed with me: "I can do all things through Christ who strengthens me" (Philippians 4:13). That verse became my mantra. Every time I felt like I couldn't go on, I reminded myself that I wasn't alone.

Finding My Village

As much as I prided myself on being independent, I eventually realized that I couldn't do it all alone. Slowly, I started to build a support system. I found other military moms who understood the unique challenges I was facing.

We swapped stories, advice, and encouragement. We became each other's lifelines, stepping in to babysit during emergencies or simply offering a listening ear. These women reminded me that I wasn't alone, and their support was invaluable.

The Lessons I Learned:

1. Prioritize What Truly Matters

In the Navy, it's easy to get caught up in the need to be the best. But motherhood taught me to redefine success. Instead of focusing on being the perfect sailor, wife, or mom, I focused on being present. My son didn't need perfection—he needed me.

2. It's Okay to Struggle

For a long time, I felt like I had to keep up a strong front. But I learned that it's okay to admit when you're struggling. Vulnerability isn't a weakness; it's a sign of strength.

3. Celebrate Small Wins

In the military, we're trained to aim for big accomplishments. But as a mom, I learned to celebrate the small victories—like getting my baby to daycare on time or making it through a long day without breaking down.

4. Lean on Your Faith and Your Community

Faith and community became my lifelines. Whenever I felt like I couldn't go on, I turned to God and the amazing people around me.

Why It Was All Worth It

Looking back, I can see how those hard days shaped me. They made me stronger, more resilient, and more compassionate. Today, when I look at my son, I see the reason for all my sacrifices. He's healthy, happy, and thriving—and he knows his mom is a fighter.

Balancing military life and motherhood is not for the faint of heart. It requires sacrifice, strength, and an endless well of love. But it's also a journey filled with unmatched pride and joy.

A Message to Fellow Moms in Uniform

To every mom out there juggling a military career and motherhood: You are a warrior. I see your sacrifices, your sleepless nights, and your unwavering determination. Remember that you are not alone. Lean on your faith, your community, and your inner strength.

Most importantly, give yourself grace. You're doing the best you can, and that is more than enough. Whether you're in uniform or out, whether your day was perfect or a total disaster, know this: you are an incredible mom, and your efforts do not go unnoticed.

Keep going. Keep fighting. You've got this!

General Austin Lloyd and I in
Baghdad Iraq 2011- 2012 deployment
Former United States Secretary Of Defense

Chapter 7: Life Lessons from the Battlefield

The Battlefield of Iraq

The journey began in Kuwait, where I underwent intensive training in preparation for my deployment to Iraq. It was here that I boarded a flight into Baghdad, filled with equal parts fear and determination. My assignment placed me with an elite unit, the Joint Force Special Operations Team, better known as JFSOC-I. At the time, I had no idea how significant this role would become or how much it would push me beyond my perceived limits.

Working with JFSOC-I meant stepping into a world where every decision carried immense weight. Little did I know that I would be traveling and flying across Iraq, often into dangerous and unpredictable situations. Each mission was both terrifying and courageous, requiring a level of focus and resolve I never imagined I possessed.

Closing Down Iraq: A Mission of Historic Importance

One of the most challenging yet rewarding tasks of my Naval career was being part of the effort to close down U.S. operations in Iraq. This monumental mission required relocating our team multiple times, from Baghdad to the International Zone, and later to Al Asad Airbase. Each move brought new challenges—coordinating logistics, ensuring the safety of personnel, and adapting to ever-changing conditions.

Flying across Iraq was a nerve-wracking experience. The sound of the rotors, the weight of the gear, and the knowledge of the risks ahead made every flight a test of courage. Yet, in those moments, I reminded myself why we were there: to bring stability to a region in turmoil and ensure the mission's success.

The Hardest Job, the Greatest Reward

Serving with JFSOC-I was the most demanding assignment of my Naval career, but also the most fulfilling. It required me to operate under immense pressure, make split-second decisions, and trust my instincts completely. I grew not only as a service member but also as a leader.

What made this mission so impactful was its significance. We weren't just performing routine duties; we were closing a chapter of history, laying the groundwork for the future of Iraq, and ensuring the safe return of countless service members. To know that I played a role in such a monumental effort fills me with pride to this day.

Courage in the Face of Fear

Every mission, every flight, and every relocation forced me to confront my fears head-on. Whether flying into areas with active threats or navigating the complexities of joint operations, there was no room for hesitation. Fear was always present, but it became a motivator rather than a deterrent.

This courage was not just mine—it was shared by every member of my team. Together, we faced the unknown with unwavering resolve, relying on each other to complete the mission and make it home.

Lessons for a Lifetime

The lessons I learned during my time with JFSOC-I are ones I carry with me to this day. They've shaped my approach to leadership, business, and life. I learned the value of adaptability, the strength of teamwork, and the importance of perseverance.

As I transitioned to civilian life and built Boss Lady Bling Blingy, these lessons became the foundation of my success. Whether designing a custom piece for a client or mentoring a fellow veteran, I draw on the resilience and determination forged during my time in Iraq.

Final Thoughts

Serving in Iraq with JFSOC-I was a defining chapter in my life. It pushed me to my limits, tested my courage, and taught me lessons that continue to guide me. It was the hardest job I've ever had, but also the most rewarding.

Looking back, I am grateful for the experience and the incredible people I served alongside. Together, we faced challenges most could never imagine and emerged stronger for it. My journey from those missions to where I am today is a testament to the power of resilience, teamwork, and faith in the mission.

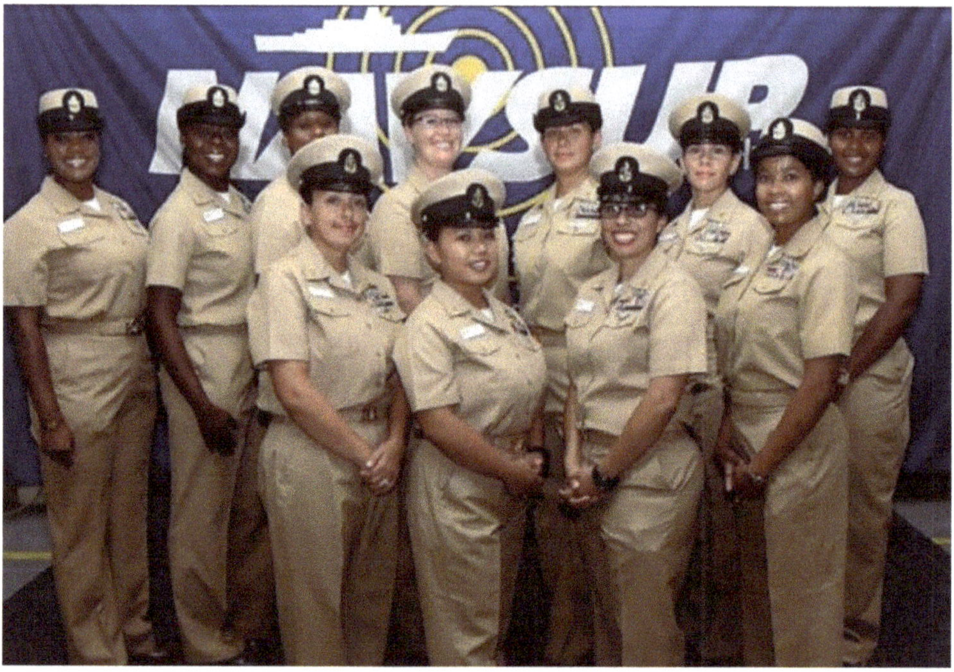
NAVSUP

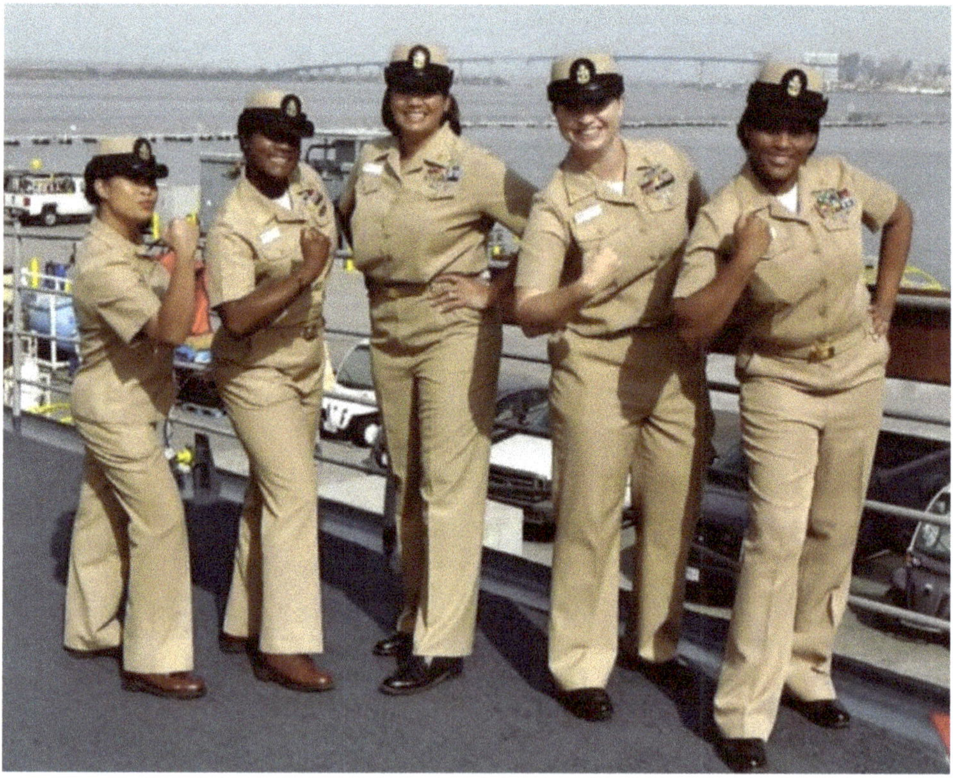

Chapter 8: Facing Adversity

The Toughest Battles Aren't Always Fought Overseas

As time went by and I went up in rank there were many challenges such as sexism, racial discrimination, or internal struggles within the military system. Now as a Senior Chief Petty Officer in the United States Navy, standing at the crossroads of history, tradition, and transformation, I have been privileged to serve my country in ways that both challenge and inspire me daily. Yet, my journey has been shaped not only by the demands of military duty but also by my unique identity as a Black woman. From the moment I first donned the uniform, I understood that my presence in this space—at this rank—was a testament to the resilience of countless women who came before me. However, it also carried the weight of the challenges and barriers that women like me continue to face in the military today.

I stand before you to recognize and celebrate the invaluable contributions of women in the military. For centuries, women have been the unsung heroes of warfare—nurses, medics, support staff, and, more often than not, the backbone of families left behind. I would like to honor our roles not just in support, but on the front lines, in leadership, and in every aspect of military service.

The Path to Equality: A Legacy of Struggle and Triumph

The story of women in the military is one of incredible perseverance. In 1948, with the passage of the Women's Armed Services Integration Act, women were officially allowed to serve as permanent, regular members of the armed forces. This was a groundbreaking moment, but it was just the beginning of a long journey toward equality. Since then, women have shattered glass ceilings, pushing the boundaries of what was once thought possible. They have taken on roles as pilots, combat engineers, special operations forces, and commanders of entire fleets and divisions. They have proven, time and again, that courage, skill, and leadership are not defined by gender but by commitment, perseverance, and the willingness to serve.

However, this journey has not been without its challenges. For women of color, particularly Black women, the path has been doubly difficult. We have fought not only to be recognized for our leadership potential but also to confront the deep-seated biases that exist within our ranks. Discrimination, both overt and subtle, has been an enduring reality for many of us. Yet, we have persisted. We have proven that we are not defined by our race, our gender, or the limitations others may place upon us. We are defined by our actions, our integrity, and our commitment to the mission.

The Unique Challenges Faced by Women of Color in the Military

As a Black woman in the military, my journey has been shaped by the intersections of race and gender. From my first day in uniform, I was acutely aware of the fact that I was entering a space that was not designed with someone like me in mind. The military has historically been an institution where the dominant culture has been white and male. Navigating this environment has meant continuously challenging stereotypes and disproving misconceptions about my abilities and my place in the ranks.

One of the unique challenges that many women of color face in the military is the pressure to represent not just ourselves but our entire race and gender. We are often seen as "the exception," held to higher standards than our counterparts and scrutinized more closely. While this can be an exhausting burden to bear, it also serves as a driving force. We understand that every promotion, every achievement, is not just a reflection of our individual efforts but a step forward for all who will come after us.

At the same time, the military has provided me with an opportunity to demonstrate the depth and breadth of my capabilities. My experiences in leadership, mentoring, and problem-solving have proven that race and gender are not barriers to success. I have had the honor of serving alongside some of the most talented and dedicated individuals in the world—both men and women—and I've learned that the key to success is not what we look like on the outside but what we bring to the table.

The Role of Women in Military Leadership

Leadership in the military, as in any other institution, is about inspiring and guiding others toward a common purpose. As a Senior Chief Petty Officer, I have learned the importance of not only leading from the front but also mentoring the next generation of leaders. In this role, I have had the privilege of working with a diverse group of men and women, each with their own backgrounds, strengths, and challenges. This diversity has enriched my leadership style, encouraging me to take a more inclusive and collaborative approach to problem-solving and decision-making.

Women in leadership positions in the military have brought a unique perspective that has proven invaluable in navigating complex situations. We often emphasize diplomacy, collaboration, and empathy—qualities that have enhanced the military's ability to respond to a range of situations, from conflict zones to peacekeeping operations. This leadership style has made me a better leader, and it has also helped to foster a culture of respect and understanding among my colleagues.

The presence of women in leadership roles has helped to redefine what it means to be strong in the military. Strength is no longer defined solely by physical power or the ability to engage in combat. Strength also comes from resilience, adaptability, and the capacity to support and uplift others. As women in leadership positions continue to rise, the military becomes stronger, more diverse, and better equipped to face the challenges of the modern world.

Shifting the Culture: Diversity and Innovation in the Military

As more women, and women of color, rise through the ranks, the culture of the military is evolving. The inclusion of women has fostered greater diversity, leading to more innovative thinking and problem-solving. Military leaders have come to recognize that strength, strategy, and leadership do not have a gender. In fact, the inclusion of women has only strengthened our armed forces, making them more reflective of the society they protect. This diversity has led to better decision-making and more effective outcomes, as we bring together a wide range of perspectives and experiences to tackle complex challenges.

The importance of diversity in the military cannot be overstated. By encouraging people from all walks of life to serve, the military ensures that it draws from the best and brightest individuals—regardless of race, gender, or background. Women of color, in particular, have brought a unique blend of resilience, creativity, and strength to the table. We understand the importance of teamwork, perseverance, and adapting to changing circumstances. These qualities make us excellent leaders and invaluable members of the military community.

The Future: Continuing the Fight for Equality

Despite the strides we have made, the work is far from over. Women in the military continue to face challenges that require concerted efforts to address. From combating harassment and discrimination to ensuring that women have equal opportunities for advancement, there is still much to be done. As a Senior Chief Petty Officer, I am committed to advocating for a military culture that fully supports women, ensuring that they have the resources and opportunities they need to thrive.

We must continue to push for a military that values diversity not just in numbers but in action. This includes providing

mentorship and support to women who aspire to reach higher ranks, offering flexible policies for those balancing family life with military service, and creating a culture where everyone, regardless of their background, feels empowered to succeed.

Honoring the Legacy and the Future

Today, I honor the women who have served in the military, I also look ahead to the future and celebrate the trailblazers of the past, to all the women who fought for the right to serve, and the women who continue to break barriers every day. But I also honor those women who will answer the call tomorrow, ready to lead, to fight, and to serve with courage and determination.

As a Black woman and a Senior Chief Petty Officer in the United States Navy, I am proud of the progress we have made and the obstacles we have overcome. But I also recognize that our journey is not over. We will continue to push for greater inclusion, equity, and respect. We will continue to break down barriers, inspire future generations, and prove that strength, leadership, and courage are not defined by gender or race, but by the commitment to serve and protect our nation.

Let us honor the women who have come before us, the women who serve with us today, and the women who will continue to forge new paths for generations to come. They are leaders, warriors, and patriots, and they deserve our utmost respect and admiration.

Chapter 9: The Decision to Retire

Saying goodbye to the Uniform.

After 25 years and two months of service, I made one of the hardest decisions of my life—to retire from the United States Navy. For over two decades, the Navy wasn't just my career; it was my identity. The uniform symbolized strength, purpose, and leadership. But beneath the surface, I was crumbling.

My personal life was falling apart. After 22 years of marriage, my husband decided he no longer wanted to be married. The words hit me like a ton of bricks. We had built a life together, shared dreams, and raised our son, who had recently graduated from college. Suddenly, the foundation I thought was unshakable cracked beneath me. The house was empty, the silence was deafening, and the sense of loss was overwhelming.

I tried to hold it together, but my world was spinning out of control. At work, I put on a brave face, leading Sailors, and making mission-critical decisions. But inside, I felt like a fraud. How could I lead others when I could barely keep myself together? How could I inspire when I felt broken?

The stress took a toll, both mentally and physically. I lost over sixty-three pounds—not through dieting, but because I couldn't eat, couldn't sleep, and couldn't stop the whirlwind of anxiety that consumed me. Looking in the mirror, I didn't recognize myself. The woman staring back was a shadow of who I used to be. I was exhausted, drained, and depressed.

Even the things I loved—things that once brought me joy—felt meaningless.

I battled with the decision to retire for over a year. I beat myself up constantly, replaying "what if" scenarios in my mind. What if I stay and things get better? What if I leave and regret it? The indecision was paralyzing, but I knew deep down that I couldn't go back out to sea in this condition. I couldn't lead others effectively when I was so lost myself.

The weight of it all finally broke me. One sleepless night, sitting alone in the quiet of my home, I hit my lowest point. I asked myself the hardest question of all: How did I get here? How did a woman like me go from having it all to losing it in just a few months?

And then, clarity struck. This is it; I told myself. I can't keep living like this. I had to let go—not just of the uniform, but of the guilt, the shame, and the unrealistic expectations I'd placed on myself for so long. That night, I made the decision to retire.

It wasn't an easy choice. Walking away from a career that had been my life for 25 years felt like stepping off a cliff into the unknown. But it was necessary. I realized that to heal, I had to stop clinging to the past and start building a future that prioritized my well-being.

Retirement was the beginning of a new chapter—one filled with uncertainty but also with possibility. I started to rediscover who I was outside of the Navy. I learned to embrace my vulnerability and accept that it's okay not to be okay. Slowly but surely, I began to rebuild.

I poured my energy into my passions: designing jewelry, empowering others, and sharing my story. Each step forward brought me closer to the woman I was meant to be—strong, resilient, and unapologetically me. Letting go of the uniform wasn't just about leaving the Navy; it was about reclaiming my life.

This journey hasn't been easy, and there are still days when I miss the camaraderie, the purpose, and the structure that the Navy gave me. But I've learned that it's okay to close one chapter to open another. I've learned that my value isn't tied to a uniform but to the impact I make and the lives I touch.

If you're struggling with a similar decision, know this: it's okay to let go. It's okay to prioritize yourself and step into the unknown. Because sometimes, the hardest decisions lead to the most beautiful transformations.

Please see Press Release:

** Logistics Specialist Senior Chief Petty Officer (SW/AW/SCW) Latasha Fennell Press Release **

FOR IMMEDIATE RELEASE.

Senior Chief Petty Officer Latasha Fennell Retires After 25 Years of Transformative Service and Leadership in the United States Navy.

The United States Navy proudly announces the retirement of Senior Chief Petty Officer Latasha Fennell after 25 years of unapparelled service, distinguished leadership, and groundbreaking contributions to the nation and her

fellow Sailors. Fennell, known for her tireless work ethic, compassionate mentorship, and commitment to excellence, leaves a legacy that stands as a testament to the power of service, dedication, and resilience.

Fennell's Navy journey began in 1999, shortly after graduating high school. Inspired by a desire to see the world and serve her country, Fennell entered the Navy as a Storekeeper, a role later redefined as Logistics Specialist to better reflect the evolving demands of military and civilian logistics. From her first deployment to Guam, where she experienced the wonders of overseas service, to subsequent missions spanning Asia, Australia, and beyond, Fennell dedicated herself to every assignment with pride and vigor. Fennell spoke of her joy during her service stating that "Serving in the Unites Navy transformed me. I was a country girl looking to see the world and now I have tangibly contributed to one of the greatest nations of all time. The pride and honor I feel cannot be expressed."

In each position, Fennell exemplified the Navy's values, quickly progressing through the ranks. Fennell's discipline, commitment and courage out shone many enabling her to reach the esteemed rank of Senior Chief Petty Officer (E-8), a level below the highest enlisted rank. Marked by her leadership and expertise, Fennell demonstrated invaluable contributions across multiple commands, with responsibilities ranging from managing logistics for over 5,700 personnel on aircraft carriers to overseeing multi-million-dollar aviation parts.

Throughout her career, Fennell brought compassion and humanity to her military service, forging deep connections, and creating a supportive environment for her fellow Sailors. "You brought joy into the squadron every single day," said Captain Richard Haley, USN. "I was constantly thinking of ways to take care of our Sailors, and you were beating me to the punch. You personally organized potlucks to feed our Sailors or baking and cooking for them, you were the first to know when Sailors were struggling with personal issues, always finding ways to help them through it."

Captain Haley speaks of Fennell's nobility, which resonated deeply with those who served alongside her. Fennell's initiatives within the squadron became a hallmark of her service, reflecting her commitment to creating an inclusive, caring, and resilient Navy family.

Fennell's extensive career includes a year-long deployment in Baghdad, Iraq, where she served with the Joint Special Operations Command (JSOC). In a unit where she was the only Navy personnel, Fennell was tasked with managing crucial supplies, weapons, ammunition, and logistics. Fennell ensured that her unit had the resources to accomplish their varied missions and her role in this high-stakes environment not only demonstrated her logistical prowess but also her courage and resilience. "I am one of the lucky ones to come back with ten toes and ten fingers. Serving in such a high-stakes environment solidified my commitment to making a difference, and I'm grateful to have returned safely," Fennell remarked.

Among her numerous accolades, Fennell earned the Navy and Marine Corps Achievement Medal (8), Good Conduct Medal (8), and the Joint Service Commendation Medal. Her achievements include certifications as a Surface Warfare Specialist, Aviation Warfare Specialist, and Seabee Combat Warfare Specialist, demonstrating her commitment to excellence and specialization within the Navy. Beyond these military commendations, Fennell received the President Volunteer Service Award (Gold) and was named the Global ICONIC Changemaker of the 21st Century.

Fennell's extensive training is equally remarkable, including advanced courses such as the Senior Enlisted Academy, Warrior Skills Instructor, and Financial Management Operating Target Account Leader. She has also served as an instructor, sharing her knowledge as a Professional Development Instructor and CMD Navy Leader Development Facilitator, further amplifying her positive impact on the Navy community.

Throughout her career, Fennell's commitment to service extended beyond the Navy. She actively engaged in volunteer work, earning multiple Humanitarian Awards and the prestigious California Kindness Award. Her impact on the local and global community has been acknowledged with certificates from various state and national leaders, including Congressman Jay Obernolte and the State of California.

Her journey reflects the unique challenges and opportunities faced by women in the military, a perspective Fennell champions with resilience and pride. "Most of the roles I've held aren't traditionally ladylike," Fennell noted. "Being a Black woman in combat units, such as Construction Mobile and Amphibious Battalions, meant constantly proving myself and showing that women can excel in these demanding fields. Women do amazing things, and my hope is to inspire others to pursue their dreams fearlessly."

As part of her service, Fennell also took on the demanding role of a recruiter, managing recruitment efforts across Louisiana and Florida. Her leadership during Hurricane Katrina earned her and her team multiple awards, including Station of the Year and the Recruiting Gold Wreath (5), showcasing her adaptability and ability to inspire those around her.

With her retirement, Fennell is charting a new course as an entrepreneur and advocate for female empowerment. Drawing on her love of fashion and resilience, she launched "Boss Lady Blingy," a luxury brand dedicated to empowering women through fashion. "Boss Lady Blingy represents strength, resilience, and the bold commitment to being unapologetically yourself," Fennell explained. "It's about standing out, feeling good, and embracing who you are. My hope is to show other women that if I can do it, they can too."

The brand's commitment to luxury and style reflects Fennell's dedication to quality and empowerment. Through custom, bling-forward designs, her creations celebrate individuality and inspire confidence, encouraging women to stand out in any setting. "This brand is not just about looking good; it's about helping people feel good and giving them a way to express themselves authentically," Fennell said. Her business also incorporates unisex designs, offering options for everyone while staying true to her empowering ethos.

As Fennell transitions from military to civilian life, she carries forward her motto: "You can't fly without supply." This phrase served as her bedrock throughout her career and underscores her belief in the power of logistics and the essential role of supply and demand in both military and civilian life. Her journey, marked by courage, service, and an unwavering commitment to helping others, serves as an inspiration to future generations of service members and civilians alike.

Fennell's legacy in the Navy is not only defined by her professional achievements but also by the lives she touched and the positive impact she made on those around her. As she begins this new chapter, she remains dedicated to her mission of inspiring women and leading by example, proving that with strength, resilience, and a commitment to one's values, anything is possible.

BOSS LADY
BLING BLINGY

For more information or media inquiries, please contact:

Amb. Dr. Latasha Fennell (h.c.)
Ret. Navy Senior Chief Petty Officer
Boss Lady Bling Blingy Boutique
2031 Commercial St San Diego Ca 92113
Tel: 619-617-4586
Website: BossLadyBlingBlingy.com
Email address: latashafennell@gmail.com

BOSS LADY
BLING BLINGY

Chapter 10: Adjusting to Civilian Life: A New Mission

A New Mission as Boss Lady Bling Blingy

After 25 years of dedicated service in the United States Navy, I transitioned into civilian life—a journey I often describe as both liberating and disorienting. Accustomed to the structured, mission-driven world of the military, I found myself navigating a life where the routine and purpose I'd known for decades were no longer prescribed.

The early days of my retirement were marked by reflection and uncertainty. "When you take off the uniform, it's like shedding a piece of your identity, I knew I had a new mission ahead, but I wasn't sure what it was yet." While I celebrated my freedom to explore new passions, there was an undeniable void—a need to redefine my purpose.

Challenges of Civilian Life

The first hurdle was adapting to a slower pace. In the Navy, Latasha thrived on discipline and deadlines, often juggling multiple high-stakes tasks. Civilian life felt, in her words, "unstructured and quiet." It was a stark contrast to the camaraderie and constant challenges of military service.

Another challenge was translating my military experience into the civilian world. While my leadership skills and resilience were undeniable, I faced moments of doubt about how they would fit into my new reality. "You spend years as part of a team with a shared mission, and suddenly, you're a team of one trying to figure out what's next,

I also grappled with the emotional toll of leaving a close-knit community. The bonds forged in the Navy were irreplaceable, and transitioning to a world where those connections weren't built into my daily life was daunting.

Finding a New Mission

The turning point came when I rediscovered a childhood passion—design I have always been known for my creativity and eye for detail, and I began experimenting with jewelry-making as a therapeutic outlet. What started as a hobby soon blossomed into a vision: Boss Lady Bling Blingy, a boutique celebrating individuality and empowerment through custom jewelry, fashion, and accessories.

I have poured her heart into my new mission, channeling the discipline and determination honed in the Navy into my entrepreneurial journey. The boutique became more than a business; it was a platform to uplift others. Whether designing a custom blazer for a military retirement or crafting a vintage-inspired necklace, I found joy in helping clients shine.

I also used my platform to give back, donating a portion of my proceeds to Veteran Disability Services (VDS). "Serving those who've served is my way of staying connected to the military community,

Embracing the New Mission

Through Boss Lady Bling Blingy, I embrace a dual role as an artist and advocate. My boutique has become a hub of creativity and empowerment, reflecting my values of resilience, individuality, and service.

This journey wasn't without its challenges—learning to run a business, balancing motherhood, and navigating a new identity—but it was a mission I approached with the same courage and commitment as my military career.

My story is a testament to the power of reinvention. By embracing my creativity and staying true to my values, I have found a new sense of purpose. "Adjusting to civilian life wasn't easy but through Boss Lady Bling Blingy, I've discovered that my mission to serve and inspire didn't end when I retired—it just evolved."

My journey from combat boots to blingy high heels is not just about fashion; it's about resilience, transformation, and finding purpose in unexpected places.

Chapter 11: Fashion and Finding my Purpose

Fashion has always been more than just clothes and accessories for me—it's been my language, my expression, and ultimately, my purpose. Even as a young girl, I was drawn to the world of glitter, style, and creativity. My journey into the world of fashion and design, however, truly began after I stepped out of my United States Navy uniform for the last time, following 25 years of dedicated service.

Looking back, I realize my love for handcrafted jewelry and all things blingy, began long before I even understood what it meant to follow your passion. One of my fondest memories is of my father giving me Ring Pops and those colorful sugar candy necklaces. I didn't just eat them; I wore them, relishing the sparkle and color they brought to my little world. Even then, I knew jewelry wasn't just about adornment—it was a statement, an identity.

As a child, I was captivated by the women in my family. I'd watch my mother, aunties, and older cousins prepare for a night out. Their rituals of getting dressed—the careful selection of outfits, the styling of their hair, the application of makeup, and the final touch of fancy shoes—felt like a magical transformation to me. I'd sit in awe, studying every detail, dreaming of the day when I could create my own looks and shine just like them.

Life took me on a different path at first. The Navy taught me discipline, resilience, and leadership, but it also brought out my innate creativity. I loved finding ways to stand out even in uniform—whether it was a carefully chosen accessory or a polished appearance, I always found a way to infuse a bit of "Latasha flair."

When I retired, I faced the daunting question: What now? While many veterans struggle to find their footing after service, I leaned into my lifelong passion. Fashion and jewelry became my outlet—a way to rediscover myself and redefine my purpose.

Creating jewelry and designing custom pieces didn't feel like work—it felt like home. I poured my heart into crafting one-of-a-kind items that sparkled with love and personality. My designs were not just about aesthetics; they told stories and celebrated individuality. Soon, my hobby turned into Boss Lady Bling Blingy Boutique, where my mission is to empower others to feel confident, bold, and radiant.

Reflecting on my journey, I see how every step prepared me for this moment. The discipline from my military years, the inspiration from my family, and my childhood love for anything blingy have all come together to shape the woman I am today. I've traded my combat boots for blingy high heels, but my mission remains the same: to inspire, empower, and shine.

Fashion isn't just about looking good—it's about discovering who you are and owning it. For me, it's also about creating a legacy of beauty, resilience, and purpose, one handcrafted piece at a time.

BOSS LADY
BLING BLINGY

Chapter 12: The Sparkle in Adversity: The Birth of Boss Lady Bling Blingy

Boss Lady Bling Blingy wasn't just a brand—it was the manifestation of a dream rooted in passion, creativity, and determination. It's story began in January 2018, while I was deployed to Japan. Amid the demanding routine of military life, I found solace in creating something beautiful with my hands. Making jewelry became my escape and, eventually, my purpose.

As a Chief Petty Officer in the Navy, I craved jewelry that reflected my rank and identity. I started by crafting necklaces, bracelets, and earrings featuring the Navy anchor—pieces that honored my role and symbolized strength. My designs quickly gained attention from my peers in the Chief Mess, who admired their uniqueness and requested custom pieces of their own. Their encouragement sparked the realization that this wasn't just a hobby—it was the beginning of something much bigger.

The Leap from Passion to Brand

When I returned home, my creativity and ambition outgrew the guest room and garage, which were packed with materials and finished designs. I decided to take a leap of faith and test my creations at a local farmers market in Imperial Beach. Although the turnout was underwhelming—I made only one sale—it didn't deter me. Instead, it fueled my desire to learn and improve.

I enrolled in SBA classes to master marketing and branding and began seeking out more strategic venues. Over time, I transitioned to showcasing my designs at upscale hotels and artisan markets. This time, I was ready. Customers were drawn to my vibrant displays, and my brand, Boss Lady Bling Blingy, started building a loyal following.

A Defining Moment: The House of Blues

One of the most defining moments in my journey came when RAW, a Los Angeles-based company, invited me to represent San Diego at the iconic House of Blues. It was an extraordinary opportunity, and I poured my heart into preparing for the event. That night, my table radiated with the sparkle and personality of my designs. The response was overwhelming. Guests admired my work, engaged with my story, and celebrated the creativity behind each piece. It was a total hit and marked a turning point for Boss Lady Bling Blingy.

The Sparkle Spreads

From that moment on, my brand flourished. Invitations to participate in fashion shows, museum exhibits, and special events started coming in. Boss Lady Bling Blingy became synonymous with bold, one-of-a-kind designs that exuded confidence and individuality.

Eventually, I secured a permanent home for my boutique at the San Diego Made Factory. This space wasn't just a store—it was a testament to my journey, a sanctuary for creativity, and a hub for the community that believed in me.

Lessons in Resilience

My journey with Boss Lady Bling Blingy has been far from easy. From humbling beginnings at the farmers market to weathering a global pandemic, each challenge tested my resilience. But I adapted, learned, and persevered. The House of Blues event reminded me of what was possible when I trusted in my vision and never gave up.

A Legacy of Creativity

Boss Lady Bling Blingy is more than just jewelry. It's a reflection of my story—a Navy Chief turned entrepreneur, guided by a love for sparkle and a belief in the power of creativity. Every piece I design tells a story and brings a little light into the lives of those who wear it.

Today, as I look back on my journey, I feel immense pride. Boss Lady Bling Blingy isn't just a business—it's a legacy, proof that with determination and a touch of sparkle, dreams can become reality.

Chapter 13: Building a Brand: From Vision to Reality

Starting a brand from scratch can feel overwhelming, especially when you're learning as you go. Trust me—I've been there. When I began Boss Lady Bling Blingy, I had no idea what I was doing. There were mistakes, challenges, and moments of doubt. But with persistence, resourcefulness, and lessons from my military career, I turned my vision into a thriving business. Here are the steps that helped me build my brand from the ground up:

1. Start Small and Be Resourceful

I used a bootstrapping system to fund my business. Each month, I set aside a portion of my income to invest in materials and products. Whether it was jewelry supplies, custom design tools, or marketing materials, I prioritized small, steady investments that aligned with my goals. Starting small taught me to maximize every dollar and appreciate the power of resourcefulness.

2. Leverage Your Skills and Experience

My military background as a Logistics Specialist Senior Chief Petty Officer gave me tools that became invaluable as an entrepreneur—organization, leadership, and problem-solving. Whatever your background, take stock of the skills you already

have and apply them to your business. Your unique experiences are often the foundation of your brand.

3. Create Visibility

One of the first things I focused on was showcasing my work. Every month, I created or purchased marketing items that told the world who I was and what I offered. Whether it was business cards, a social media presence, or product displays, I made sure my brand had visibility.

4. Educate Yourself

I attended every business and entrepreneurship class the Navy offered. Knowledge is power, and the more I learned about running a business—whether it was finance, marketing, or networking—the more confident I became. Don't hesitate to seek out free or affordable resources to sharpen your skills.

5. Build Relationships and Network

I showed up. Markets, pop-ups, local events—you name it, I was there. Selling my products at local markets like the Imperial Beach Park Market wasn't just about making sales; it was about meeting people, building relationships, and learning from other entrepreneurs. Networking is key to growing a brand.

6. Join Communities

I joined entrepreneur groups and organizations that provided support and inspiration. Surrounding yourself with like-minded people is crucial—they can share advice, offer encouragement, and even become collaborators or customers.

7. Be Patient and Stay Consistent

Building a brand takes time. I didn't start with a store or a boutique—it was a journey of gradual growth. Every small step mattered, and consistency was my secret weapon. I stayed committed, even during slow seasons or moments of doubt.

8. Adapt and Innovate

Challenges will come, and adaptability is key. When sales were slow or circumstances changed, I pivoted. For example, during COVID-19, I focused on creating blingy masks and custom projects, which helped me stay afloat. Always look for ways to innovate and meet your customers' current needs.

9. Stay True to Your Vision

Your brand is a reflection of you. Keep your vision at the heart of everything you do. For me, Boss Lady Bling Blingy isn't just about jewelry or fashion—it's about empowering others to feel bold, confident, and unique.

10. Celebrate Every Milestone

No matter how small, every achievement is worth celebrating. Whether it's making your first sale, joining a new market, or completing a challenging project, these moments fuel your motivation and remind you why you started.

Building a brand is a marathon, not a sprint. It requires passion, persistence, and patience. If I can do it, starting with no prior experience and learning along the way, so can you. Remember, every step you take brings you closer to turning your vision into reality.

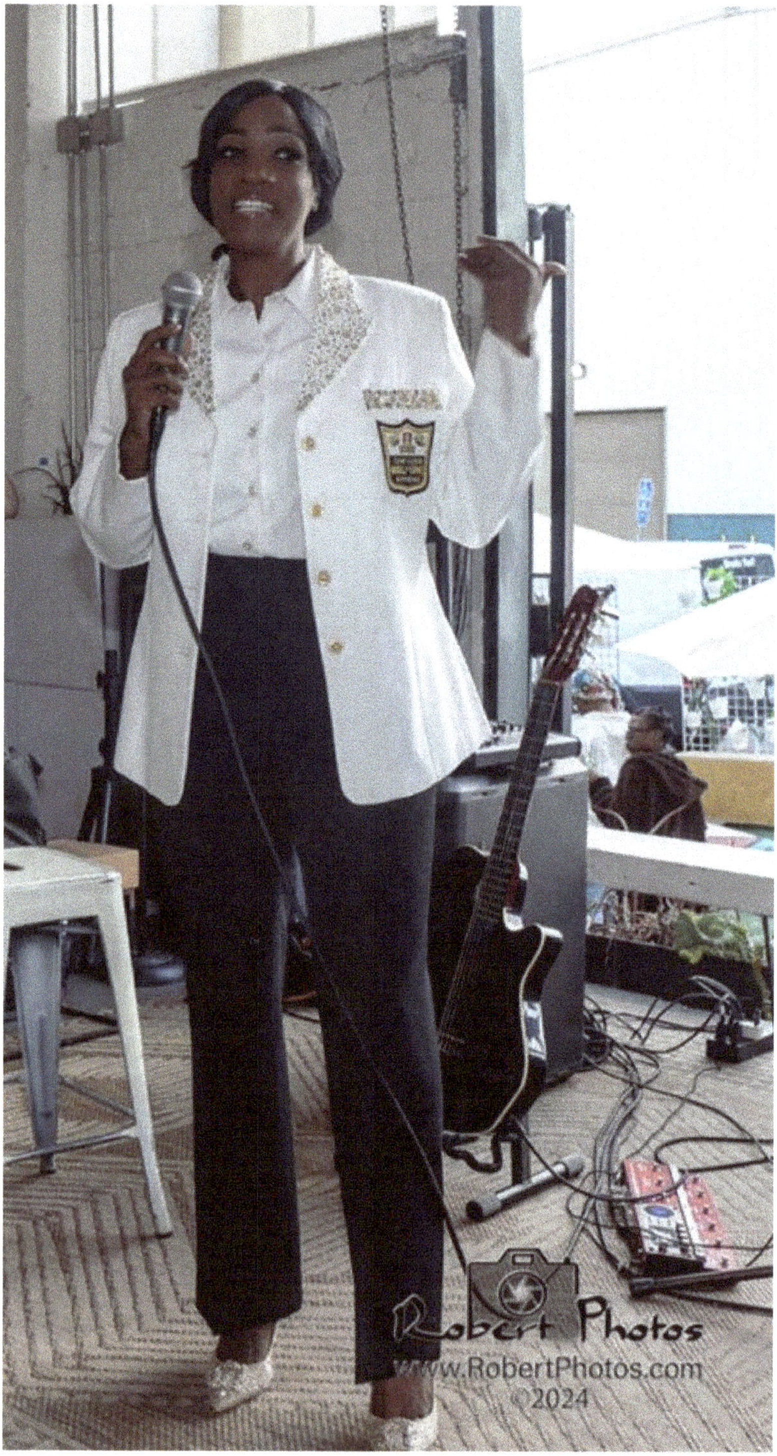
Robert Photos
www.RobertPhotos.com
©2024

Chapter 14: Finding My Voice

From Veteran to Businesswoman

Transitioning from a structured military career to the creative and competitive world of business was one of the most challenging and rewarding journeys of my life. As a retired United States Navy Senior Chief Petty Officer, I was accustomed to order, discipline, and teamwork. But stepping into the civilian business world was like entering uncharted waters—cutthroat, unpredictable, and lacking the structure I had always thrived in. This is my story of navigating that shift and finding my voice as a businesswoman.

The Military Mindset

In the Navy, being outspoken and heard wasn't just encouraged—it was essential. Clear communication, accountability, and leadership were part of the job, whether managing a team, overseeing logistics, or making split-second decisions in high-pressure situations. The military instilled in me a strong sense of responsibility, punctuality, and the importance of following through. Being on time wasn't just expected; it was non-negotiable. Leading people came naturally because it was what we did every day.

But as I soon discovered, civilian life operated by a completely different set of rules. While I was equipped with the skills to lead, organize, and execute, I quickly realized that civilian life required me to adjust not just my approach, but also my expectations.

The Shock of Civilian Life

One of my first challenges was learning to communicate effectively with civilians. In the military, you give an order, and it's carried out without hesitation. Everyone understands the chain of command and their role within it. In the civilian world, I encountered people who had no concept of "military bearing"—individuals who gave up easily, didn't show up, or lacked the discipline I was used to seeing.

Initially, I struggled to figure out the right language to use. How could I motivate people who weren't driven by the same sense of duty and responsibility? My early days as a recruiter helped me here. In that role, I dealt with people from all walks of life, learning how to adjust my communication style to resonate with different personalities. This experience became invaluable as I transitioned into business. I realized that instead of issuing orders, I needed to inspire, persuade, and sometimes even compromise.

Adapting to Competition

One thing that came naturally to me was handling competition. In the military, we constantly competed for promotions and recognition, pushing ourselves to outperform and stand out. That mindset served me well as I entered the business world. I viewed every challenge as an opportunity to rise above the rest, whether it was perfecting my craft, expanding my network, or finding creative ways to market my brand.

Competition in business, however, felt more personal. In the military, we competed within a structured system with clear rules and outcomes. In business, there were no rules—just an endless drive to stay relevant and outshine the competition. I had to learn quickly that not everyone plays fair, and I needed to stand firm in my values while also staying adaptable.

Building Boss Lady Bling Blingy

Starting my own business, Boss Lady Bling Blingy Boutique, was both exciting and terrifying. I knew I had the leadership skills and the determination to make it work, but I had no roadmap. I relied heavily on the skills I honed in the Navy: strategic planning, resourcefulness, and the ability to think on my feet.

In the early days, I used a bootstrapping method to fund my business, setting aside a portion of my income each month to invest in materials, tools, and marketing. I wasn't afraid to start small. I created pieces for myself and shared them with others, letting word-of-mouth and local markets help me gain traction.

I also leaned into education. The Navy offered entrepreneur classes, which I eagerly attended. I soaked up every piece of advice and applied it to my growing business. I joined local entrepreneur groups, networked at markets, and sold my creations wherever I could—Imperial Beach Park, pop-up events, and beyond. Every interaction became a chance to learn and grow.

Overcoming Challenges

The transition wasn't without its setbacks. There were moments when I questioned my abilities and whether I was cut out for the business world. Civilian life lacked the camaraderie and structure I had relied on in the military. I missed the sense of unity and shared purpose.

One of the biggest challenges came in 2020, when the COVID-19 pandemic hit. Suddenly, no one was dressing up or going out, and sales for jewelry plummeted. My business took a major hit, and I had to pivot quickly. Drawing on my resourcefulness, I began creating custom blingy masks, personalized home décor, and embellished clothing to meet the new demands of the market. Those projects not only kept my business afloat but also allowed me to explore new creative avenues.

Finding My Voice

Through all these challenges, I found my voice as a businesswoman. I realized that the same qualities that made me successful in the military—resilience, adaptability, and a commitment to excellence—could help me thrive in business. But I also had to develop new skills, like building relationships, marketing my brand, and navigating the unpredictable world of entrepreneurship.

I learned that finding your voice doesn't mean shouting louder than everyone else. It means being authentic, standing firm in your values, and staying true to your vision. For me, that vision was about empowering others to feel bold, confident, and unique through my designs.

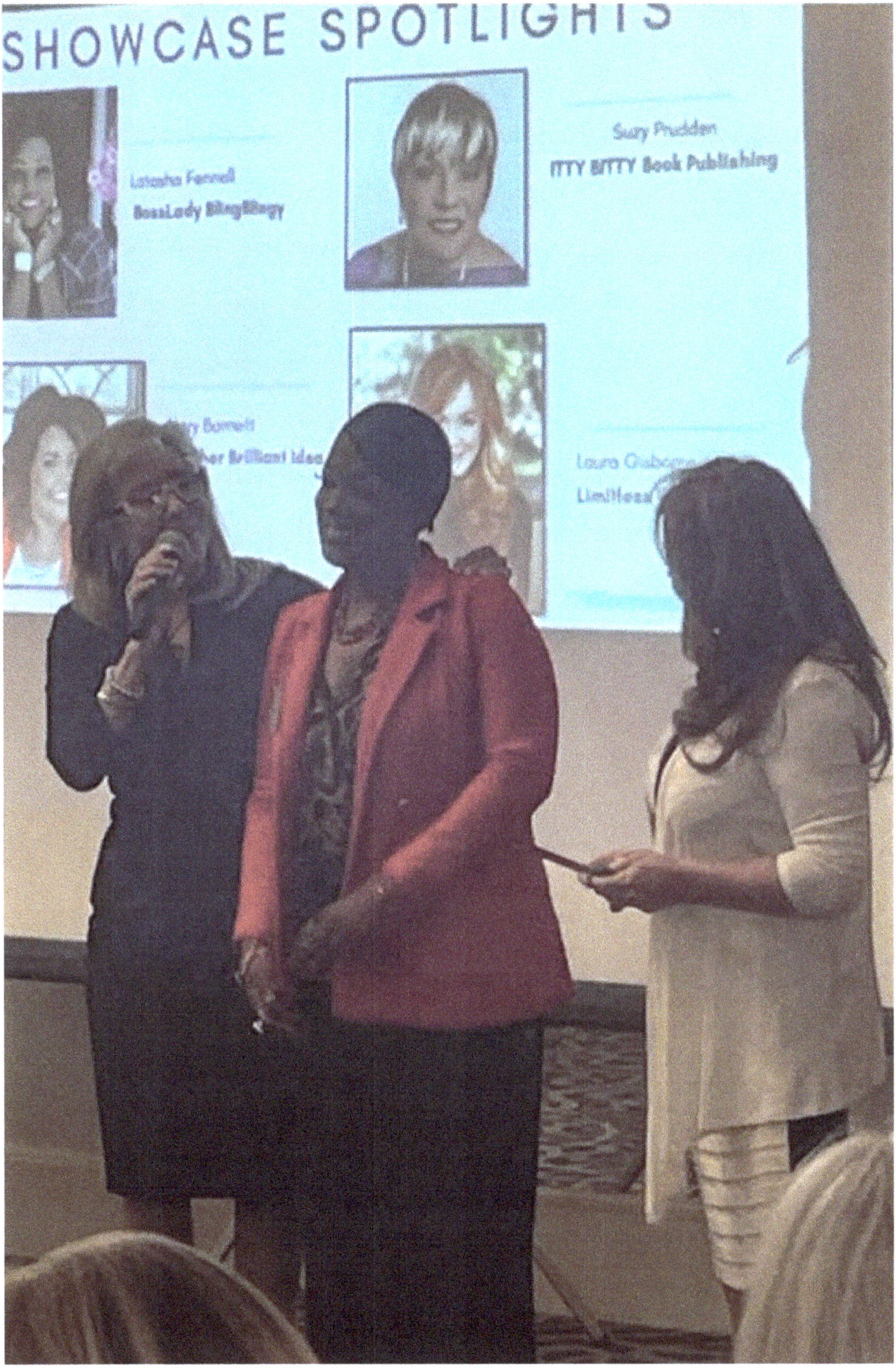
SHOWCASE SPOTLIGHTS
Latasha Fennell
Suzy Prudden
ITTY BITTY Book Publishing

The Rewards of Entrepreneurship

Today, Boss Lady Bling Blingy is more than a business—it's a reflection of who I am. It's a testament to the lessons I've learned and the challenges I've overcome. I've gone from creating jewelry for myself to crafting one-of-a-kind pieces for clients who want to stand out and shine.

The rewards of entrepreneurship go beyond financial success. It's about creating something that's uniquely yours, something that impacts others and leaves a legacy. It's about proving to yourself that you can adapt, grow, and thrive no matter what life throws your way.

Advice for Fellow Veterans and Aspiring Entrepreneurs

If you're a veteran transitioning to civilian life or considering starting your own business, here's my advice:

1. **Leverage Your Skills:** The discipline, leadership, and adaptability you gained in the military are invaluable in business. Use them to your advantage.

2. **Stay Resourceful:** Start small and build gradually. Use what you have and invest wisely.

3. **Network and Learn:** Attend classes, join groups, and surround yourself with like-minded individuals. Every connection can open new doors.

4. Be Resilient: Challenges will come, but don't give up. Adapt, pivot, and find new ways to stay relevant.

5. Find Your Voice: Stay true to your values and vision. Authenticity is your greatest asset.

Conclusion

The transition from veteran to businesswoman wasn't easy, but it was worth it. I've discovered that the skills and lessons from my military career are the foundation of my success, but finding my voice in the business world required me to grow in new ways.

Today, I'm proud to call myself not just a veteran, but also an entrepreneur, a designer, and a changemaker. My journey has taught me that no matter where you come from or what challenges you face, you can create a life and a legacy that's uniquely your own. All it takes is determination, resilience, and the courage to find your voice.

Chapter 15: Blingy High Heels

Confidence and Identity Through Fashion

Fashion is more than just clothes and accessories; it's a way to tell the world who you are without saying a word. For me, fashion and jewelry have always been my tools for self-expression, my armor, and my statement of empowerment. And there's no better symbol of confidence and identity than a pair of blingy high heels—a combination of elegance, boldness, and a touch of flair that demands attention.

This is the story of how my love for fashion, jewelry, and, yes, those sparkling, bedazzled high heels shaped my identity, boosted my confidence, and empowered me to embrace every part of myself.

A Love for Sparkle from the Start

My fascination with bling started early. As a child, I was obsessed with anything that sparkled. My dad used to bring me Ring Pops and candy necklaces, and while most kids would devour them immediately, I wore mine like prized jewelry. To me, they weren't just candy—they were accessories, proof that even at a young age, I knew the power of adornment.

At family gatherings, I'd watch my mom, aunts, and older cousins get ready for nights out. They'd put together stunning outfits, style their hair, slip on fancy shoes, and finish the look

with makeup and jewelry. Watching them transform was mesmerizing. I couldn't wait to grow up, dress up, and shine just like them.

But it wasn't just about looking good—it was about how they felt. You could see the confidence in their walk and the joy in their laughter as they strutted out the door, ready to take on the world. I realized then that fashion was more than appearances; it was a source of power.

High Heels as a Statement of Confidence

As I got older, I discovered the magic of high heels. There's something about slipping into a pair of heels that instantly changes your posture, your stride, and your mindset. High heels demand attention. They say, I've arrived, and I'm not afraid to take up space.

For me, blingy high heels took that feeling to another level. I wasn't just walking; I was sparkling with every step. The rhinestones and embellishments reflected light and drew eyes, making me feel unstoppable. Blingy heels became my signature—a way to express my personality and confidence in a way that words never could.

The Journey to Self-Expression

My journey to using fashion as a tool for self-expression wasn't always straightforward. Like many women, I went through phases of self-doubt and uncertainty about who I was. My military career added another layer of complexity. While I loved the structure and camaraderie of the Navy, the uniform didn't leave much room for individuality.

But even within the confines of military life, I found ways to let my personality shine. I'd experiment with jewelry and accessories, subtly adding elements of sparkle to my otherwise structured appearance. Those small touches reminded me that no matter where I was or what I was doing, I could still hold onto my identity.

When I transitioned out of the military, I finally had the freedom to embrace my love for fashion and jewelry fully. I poured myself into creating pieces that reflected my personality—bold, unique, and unapologetically blingy.

Bling as Empowerment

There's something undeniably empowering about fashion that makes you feel good. For me, wearing my handmade jewelry or stepping into a pair of blingy heels is like putting on armor. It's a reminder that I am strong, capable, and worthy of being seen.

Creating blingy accessories and designs for others has been one of the most fulfilling parts of my journey. I've had the privilege of working with clients who, like me, want to use fashion to express themselves and feel confident. Seeing someone light up when they wear one of my designs is a powerful reminder of why I do what I do.

Fashion isn't just about trends or labels—it's about creating a version of yourself that you're proud to show the world.

The Challenges of Standing Out

Of course, standing out isn't always easy. In the competitive world of fashion, it can feel like everyone is vying for attention. But I've always believed that authenticity is what sets you apart. My love for bling isn't a gimmick—it's a genuine reflection of who I am.

There have been times when I've faced criticism or skepticism, especially from those who don't understand the appeal of sparkle and shine. But I've learned that staying true to yourself is the ultimate act of confidence. When you love what you're wearing and feel good in it, the opinions of others fade into the background.

Creating Confidence Through Design

One of my favorite aspects of being a designer is helping others discover their confidence through fashion. Whether it's a custom pair of blingy heels, a statement necklace, or a personalized blazer, my goal is always the same: to make the wearer feel empowered.

I've worked with clients preparing for special events, military retirements, and even everyday occasions where they just wanted to feel extraordinary. Each design is a collaboration, a chance to bring their vision to life while infusing it with my signature touch of sparkle.

Fashion has the power to transform—not just how you look, but how you feel.

Blingy Heels and Identity

For me, blingy heels are more than just shoes. They're a symbol of my identity as a woman, a creator, and an entrepreneur. They represent my journey from a little girl mesmerized by her mom's fancy shoes to a businesswoman using fashion to make a statement.

Blingy heels remind me that it's okay to stand out, to take up space, and to embrace every part of who I am. They're a celebration of individuality and a testament to the power of self-expression.

Lessons Learned Through Fashion

Over the years, fashion and jewelry have taught me some invaluable lessons about confidence and identity:

1. **Be True to Yourself:** Don't be afraid to embrace your unique style. Authenticity is what makes you stand out.

2. **Confidence Comes from Within:** What you wear is a reflection of how you feel. Choose pieces that make you feel powerful and confident.

3. **Fashion is a Tool, not a Crutch:** It's not about hiding behind clothes or accessories—it's about using them to amplify who you are.

4. Empower Others: Helping others discover their confidence through fashion is one of the most rewarding parts of the journey.

Looking Ahead

As I continue to grow my brand, Boss Lady Bling Blingy, I'm excited to explore new ways to use fashion and jewelry to inspire confidence and empower others. Whether it's through a dazzling pair of heels, a custom design, or a one-of-a-kind piece of jewelry, my mission remains the same: to help people shine—inside and out.

Blingy high heels are more than just a fashion statement. They're a celebration of individuality, a source of empowerment, and a reminder that confidence begins with embracing who you are.

Chapter 16: Empowering Veterans: Life After Service

Boss Lady Bling Blingy: Empowering Veterans Life After Service

Transitioning from military to civilian life can be a challenging experience for many veterans. After dedicating years of service to our country, we often find ourselves navigating an entirely different world with new rules, expectations, and opportunities. For me, as a veteran and founder of Boss Lady Bling Blingy, I've taken my experiences and turned them into a mission: empowering veterans to embrace their new roles in civilian life while looking and feeling their best.

Understanding the Transition

After serving 25 years in the United States Navy, I understood firsthand how overwhelming the shift from a structured, disciplined military career to civilian life could be. In the military, we're provided with a sense of purpose, a clear chain of command, and a community of peers who share the same values. Civilian life, however, is less predictable and often lacks the same level of support and camaraderie.

Many veterans struggle with finding their place in this new environment. Questions like, "What do I do now?" and "How do I translate my skills into this new world?" can weigh heavily on their minds. Adding to the challenge is the need to redefine

personal identity after leaving behind the uniform that has defined them for so long.

This is where fashion, style, and self-expression can play a significant role. By helping veterans feel confident in their appearance, they can begin to rebuild their sense of self and step boldly into their new lives.

Building Confidence Through Fashion

Boss Lady Bling Blingy began as a passion for creating handmade jewelry and fashion pieces, but it quickly evolved into a platform for empowerment. For me, fashion is about more than just looking good—it's about feeling good and projecting confidence. This philosophy became the foundation of my work with veterans.

Many veterans transitioning into civilian careers face the challenge of adapting their appearance to fit their new roles. Whether they're attending job interviews, starting a business, or stepping into leadership positions, the right outfit can make all the difference. I've had the honor of working with veterans to create custom pieces that not only enhance their personal style but also give them the confidence to own their new paths.

Providing Practical Support

Empowering veterans isn't just about making them look good—it's about equipping them with the tools they need to succeed. Through Boss Lady Bling Blingy, I've worked to provide practical support in several ways:

1. Customized Blingy Blazer Jackets: These jackets are perfect for military retirees and veterans transitioning into professional roles. Designed to reflect their unique personalities while maintaining a polished and professional look, these jackets are a symbol of pride and accomplishment.

2. Workshops and Styling Sessions: I've hosted sessions focused on helping veterans understand the importance of personal presentation. From selecting the right outfit for an interview to accessorizing for a formal event, these workshops provide actionable advice that veterans can apply immediately.

3. Networking and Community Building: One of the biggest challenges for veterans is rebuilding their network outside of the military. By connecting them with local businesses, entrepreneur groups, and fashion communities, I've helped them expand their opportunities while fostering a sense of belonging.

The Power of Self-Expression

Leaving the military often means leaving behind the uniform that once symbolized service, identity, and unity. For many veterans, finding a new way to express themselves is an important step in the transition process. Fashion offers a unique avenue for self-expression, allowing veterans to explore who they are outside of the military.

Through my designs, I encourage veterans to embrace their individuality. Whether it's a bold accessory that reflects their

personality or a subtle nod to their military background, each piece tells a story. By reclaiming their voice through fashion, veterans can confidently step into their new identities.

Giving Back to the Community

A portion of Boss Lady Bling Blingy's proceeds is donated to organizations like Veteran Disability Services (VDS), which provide critical support to veterans and their families. This partnership is deeply personal to me, as it allows me to give back to the community that has given me so much.

By combining fashion with philanthropy, I'm able to create a ripple effect of empowerment. Veterans who feel confident and supported are better equipped to succeed, and their success inspires others to pursue their own goals.

Lessons Learned

Working with veterans has taught me that empowerment isn't a one-size-fits-all process. Each individual's journey is unique, shaped by their experiences and aspirations. The key is to meet them where they are, listen to their needs, and provide personalized solutions that resonate with them.

It's also a reminder that confidence is contagious. When veterans feel good about themselves, it shows in the way they carry themselves, interact with others, and approach new opportunities.

A Vision for the Future

My goal for Boss Lady Bling Blingy is to continue expanding its impact, reaching more veterans, and providing even greater levels of support. This includes offering mentorship programs, collaborating with other veteran-owned businesses, and creating new designs that celebrate the resilience and strength of the veteran community.

Fashion is a powerful tool for transformation, and I'm committed to using it to uplift those who have served our country. By helping veterans transition into civilian life with style, confidence, and a renewed sense of purpose, I hope to make a lasting difference in their lives.

Empowering veterans isn't just about helping them adjust to civilian life—it's about celebrating their achievements, honoring their service, and equipping them to thrive in their next chapter. Through Boss Lady Bling Blingy, I've found a way to combine my passion for fashion with my dedication to supporting the veteran community.

Every design, every workshop, and every conversation are opportunities to remind veterans that they are strong, capable, and deserving of success. Together, we can create a future where every veteran feels empowered to shine—both inside and out.

Chapter 17. The Power of Giving Back

How I Discovered the Power of Giving Back

When I think about my journey from military service to building my brand, Boss Lady Bling Blingy, one consistent theme has remained at the core of my life: service. Service to my country as a member of the United States Navy shaped who I am. Yet, as I transitioned into civilian life, I realized that my purpose didn't stop with leaving the military. The values I had upheld in uniform—duty, leadership, and commitment—found new meaning in giving back to the veteran community.

This realization has become one of the most fulfilling parts of my life's work. My involvement with Veteran Disability Services (VDS) has shown me the power of giving back and the importance of supporting others as they navigate the complex path of life after military service.

Finding My Purpose Beyond the Military

After serving 25 years in the Navy, I faced the same challenges many veterans encounter: What's next? The military provides a clear sense of identity and purpose. When that chapter ends, it can be difficult to know where to focus your energy.

For me, creativity became my outlet. I started crafting jewelry, designing clothes, and building a business that reflected my personality. But as rewarding as entrepreneurship was, I

still felt something was missing. I realized that my sense of fulfillment had always come from helping others—something I had done throughout my military career.

That's when I decided to channel my talents and resources toward a cause close to my heart: supporting veterans.

Discovering Veteran Disability Services (VDS)

Veteran Disability Services is an organization dedicated to assisting veterans who face physical, emotional, or mental challenges as a result of their service. VDS provides critical support, from navigating the complexities of the VA benefits system to connecting veterans with resources that improve their quality of life.

When I first learned about VDS, I immediately felt a connection to their mission. I understood the struggles many veterans face, having seen them in myself and others. The transition to civilian life is difficult enough without the added burdens of physical disabilities, post-traumatic stress, or other service-related challenges. I wanted to be a part of an organization that actively worked to make a difference in the lives of those who had served.

The Decision to Give Back

As I built Boss Lady Bling Blingy, I decided to incorporate giving back into the foundation of my business. A portion of my proceeds would go to VDS, allowing me to directly contribute to their efforts. This decision was about more than just financial support—it was a way to stay connected to my roots and ensure my work had a greater impact.

Through this partnership, I've had the opportunity to meet veterans from all walks of life. Each story I hear inspires me to continue my work, not just as a designer and entrepreneur, but as someone who understands the importance of service beyond the military.

The Power of Community

One of the most profound lessons I've learned through VDS is the power of community. The veteran experience is unique and having a network of people who understand that journey is invaluable.

At VDS events and meetings, I've witnessed veterans come together to share their stories, support one another, and celebrate their resilience. This sense of camaraderie reminded me of the bonds I formed during my time in the Navy. It's a testament to the strength and spirit of the veteran community and a reminder that we're never truly alone in our struggles.

Giving back to this community is more than an obligation—it's a privilege.

Incorporating Service into My Business

Giving back through VDS has also shaped the way I run my business. Every custom piece of jewelry or clothing I create is a reflection of the values I hold dear: resilience, individuality, and empowerment.

Many of my clients are veterans themselves, and it's incredibly

rewarding to design pieces that celebrate their service or mark important milestones in their lives. Whether it's a custom blazer for a retirement ceremony or a piece of jewelry that holds sentimental value, my work allows me to honor their stories in a meaningful way.

The Importance of Service Beyond the Military

One of the greatest lessons I've learned is that service doesn't end when you take off the uniform. In fact, some of the most impactful forms of service happen after military life.

For veterans, giving back can be a powerful way to heal, find purpose, and stay connected to the values they upheld during their service. Organizations like VDS provide an avenue for veterans to support one another and create positive change in their communities.

For me, giving back has been a way to combine my skills as a designer with my passion for helping others. It's a reminder that we all have something to offer, whether it's time, resources, or simply a listening ear.

A Personal Connection to Giving Back

My commitment to giving back is deeply personal. I've experienced the challenges of transitioning to civilian life and know how important it is to have support along the way. Through my work with VDS, I've seen firsthand how even small acts of kindness can have a ripple effect.

For example, one of my proudest moments was working with a

veteran who was struggling to feel confident in her new role as a civilian professional. Together, we designed a custom blazer that reflected her personality while maintaining a professional look. When she wore it to her first job interview, she told me she felt like a new person—and she got the job.

It's moments like these that remind me why giving back is so important.

Looking Toward the Future

As I continue to grow Boss Lady Bling Blingy, I'm committed to expanding my efforts to give back. My goal is to work with more organizations like VDS, host workshops for veterans, and create opportunities for them to express themselves through fashion and art.

I also hope to inspire others—veterans and civilians alike—to find ways to give back to their communities. Whether it's through volunteering, donating, or simply offering support, we all have the power to make a difference.

Discovering the power of giving back has been one of the most rewarding parts of my journey. Through my involvement with Veteran Disability Services, I've found a way to honor my military roots, support those who have served, and create a legacy of service that extends beyond the uniform.

At its core, Boss Lady Bling Blingy is about empowerment—helping people feel confident, strong, and capable. By giving back to the veteran community, I'm able to carry that mission forward and make a lasting impact in the lives of those who inspire me every day.

Chapter 18: Making My Mark with Lookbooks and Beyond

Visualizing success

Success doesn't come without vision, dedication, and the right people by your side. My journey toward establishing my mark as a creative force began with the magic of lookbooks— visual stories that captured my designs and showcased my artistry. It's a chapter in my life that wouldn't have been possible without the unwavering support of my big brother Johnny Jones and the creative brilliance of Sheron Jones, an extraordinary photographer and videographer.

With their help, I began documenting every intricate piece of my work. From handcrafted blingy jewelry to jaw-dropping accessories and dazzling clothing designs, we transformed creativity into tangible works of art. Each photoshoot captured the boldness of my vision, featuring some of the most stunning models who brought my designs to life. Together, we created not just one but six extraordinary lookbooks—testaments to my brand's essence, style, and innovation.

These lookbooks weren't just collections of photos; they became tools that propelled my brand into new heights. They told a story of elegance and glamour, capturing the attention of people who valued artistry and uniqueness. As my designs gained recognition, they began to appear in over twenty-five prominent magazines, cementing my work as a symbol of sophistication and bling. These publications not only highlighted my jewelry and accessories but also celebrated my journey and unique approach to fashion.

From Lookbooks to the Spotlight

With the exposure came new opportunities. I showcased my work at every networking event I attended, paving the way for meaningful connections with talented individuals. One of those connections was Kevin Brame, a talented director and videographer whose work elevated my brand in unimaginable ways. Collaborating with Kevin opened doors to commercials and even a short documentary that celebrated my story and vision.

My partnership with my brother Rally Vargas and his work with California Talk Radio Station became another avenue for growth. Rally believed in my vision and gave me a platform to showcase my brand to a wider audience. Whether it was being a guest on his radio station to share my journey or getting me commercials deals with big brands ensuring that my presence made an impact.

Styling the Stars

Behind the scenes, I found my stride as a stylist. I showed up and showed out, ensuring that everyone who wore my designs looked like they belonged on the red carpet. From fashion shows and high-end Galas to award ceremonies, proms, magazine shoots, and even fancy date nights, my work began to attract celebrities and influencers in Los Angeles, Hollywood,

Long Beach, Orange County, and San Diego. I gained clients all over the world. The blingy brilliance of my creations became synonymous with elegance and luxury.

Every event was an opportunity to highlight my brand and showcase my commitment to excellence. Styling for these occasions not only amplified my credibility but also allowed me to build lasting relationships with movers and shakers in the industry. These experiences helped shape the identity of Boss Lady Bling Blingy Boutique, transforming it into a trusted name for sophistication and glamour.

Shaping a Legacy

Creating six lookbooks wasn't just a milestone; it was a movement. It showcased my ability to turn dreams into reality, from crafting beautiful designs to working behind the scenes at events where the world's most influential people gathered. With over twenty-five magazine features under my belt, I've proven that my artistry resonates on both a personal and professional level.

These experiences made me realize that "making your mark" isn't about just showing up—it's about standing out and leaving a lasting impression. With the help of my incredible team, including my family and talented collaborators, I've shown the world that it's possible to build a brand rooted in passion and creativity. Today, Boss Lady Bling Blingy is more than a boutique; it's a legacy—a testament to hard work, resilience, and the power of bling to change lives.

This Jewelry Look Book would not have been possible without the love, support, dedication & motivation that this great team have provided me. An idea I talked about for over a year has turned into a reality and one of my dreams have been achieved. Thank You!

Special thanks to all those who contributed to this project :
My Loving Husband Paul Fennell who has been there from day 1 and has always supported me behind the scenes, every late night and early morning all the breakdowns and setups. Thank you from the bottom of my heart.
My Loving Son Trenton Fennell who often stayed up late to help pack and ship every item and supported every popup shop and every sale I ever made.
Johnny & Sheron Jones for the friendship, support, amazing photos and the invaluable assistance and untold hours you two devoted in researching ideas, issues, and anecdotes.
Raven Wilson for the friendship, supporting and teaching, and giving me the courage to step from behind the scenes to shine and do what I absolutely love.

Special Thanks to all of the beautiful models: Esme Perley @esmeperley , Selina Hall @_soulbeautiful , Courtney Fox @courtney.yvonne , Vanessa Arlt @vanessajazznfashion, Alexis Zurdo @iam.alexis.official , Amanda Gwen @amanda_gwen_crawford56 , The Amazing Runway Designer: Territa Torres @territatorresdesigns and Evelyn Barnes @buttonstobridalalterations . The Head Turning Makeup Artists: Debbie Miller @workinggoddess and Eva Gonzalez @eva_glez

This team has been utterly amazing every step of the way. Your love and support made this project a great success.

Chapter 19: "The Importance of Networking and Community"

Why Networking Matters

"When I retired from the Navy after 25 years of service, I stood at a crossroads. My life had been structured, disciplined, and directed by a clear mission, but suddenly, I faced an unfamiliar world filled with possibilities—and uncertainties. It was through networking and joining communities like the Global Society for Female Entrepreneurs and E-Women Network, that I found clarity and a sense of belonging. These connections became my compass, guiding me toward success."

The Power of Connection

"The Global Society for Female Entrepreneurs opened a world of opportunities I never knew existed. More than a platform for exchanging business cards, it became a sanctuary where I could learn, grow, and collaborate with women who shared my vision. These women weren't just colleagues—they became friends, mentors, and a support system that reminded me of the camaraderie I had in the military."

Lessons Learned Through Networking:

1. **Access to Resources:** Before I joined the Global Society for Female Entrepreneurs, I had a vision for my boutique but lacked the tools to execute it. Through their workshops, I

learned how to position my brand, pitch to investors, and develop meaningful collaborations.

2. Empowerment Through Shared Experiences: Listening to other women share their struggles and triumphs reminded me that I wasn't alone in facing challenges. These stories fueled my determination to push through my own obstacles.

3. Learning to Ask for Help: As a military leader, I was used to being self-reliant. Networking taught me the power of saying, 'I need help,' and how transformative it is to let others contribute to your journey.

1. Mentorship and Recognition: It was through the Global Society for Female Entrepreneurs that I met Amb. Lady Dr. Robbie Motter (h.c), a mentor who saw potential in me before I fully saw it in myself. Her belief in me not only helped me sharpen my vision but also gave me the courage to step into the spotlight and accept recognition for my work.

2. Giving Back: Joining E-Women Network reminded me of the importance of paying it forward. Whether it's mentoring a fellow entrepreneur or donating a portion of my business proceeds to Veteran Disability Services, I've learned that true success is measured by how much you uplift others.

3. Collaborative Opportunities: One of the most memorable collaborations came during San Diego Design Week when I showcased my brand in a solo fashion show. The encouragement and support I received from my network were instrumental in making that event a success.

Tips for Building Your Own Network

1. Be Intentional: Start by identifying groups that align with your values and goals. Whether it's a professional association or a local business group, choose spaces where you feel inspired.

2. Show Up and Engage: Attending events is only the first step. Be present, ask questions, and take the time to truly listen to others.

3. Give Before You Receive: Offer help, share your expertise, and celebrate others' successes. Networking is about building trust, and generosity is the fastest way to do that.

4. Follow Up: After meeting someone, follow up with a thank-you message or schedule a coffee chat. Relationships grow through consistent communication.

From Networking to Legacy

Networking is more than exchanging pleasantries; it's about building bridges that connect us to opportunities, friendships, and growth. From my days in combat boots to now standing tall in blingy high heels, every step of my journey has been shaped by the incredible women who believed in me, supported me, and inspired me. It is through these connections that I've built not just a business but a legacy of empowerment, proving that together, we can accomplish anything.

GSFE
Global Society for Female Entrepreneurs

Chapter 20: Mentorship and Leadership

Guiding the Next Generation

Mentorship and leadership have been integral to my personal and professional journey. From my time in the United States Navy to my transition into civilian life, these principles have remained at the core of who I am and what I strive to achieve. They are not just responsibilities or titles; they are a legacy we leave behind for those who follow in our footsteps.

Lessons in Leadership and Mentorship

During my years of service in the Navy, I learned that mentorship and leadership are not about rank or authority but about influence, service, and the ability to inspire others. In the military, people show up ready to work. They honor their commitments, arrive on time, and persevere even in challenging situations, often without expecting immediate rewards.
This mindset creates an unspoken bond—a trust that every individual will fulfill their role and go above and beyond when necessary.

When I transitioned to civilian life, I quickly realized things were different. Civilians often approached commitments with a more relaxed attitude. Some would arrive late, cancel at the last minute, or center the focus on themselves. Others would quit when faced with adversity, leaving teams and projects in disarray. At first, this was frustrating and unfamiliar to me.

Adapting and Leading in Civilian Life

I had to adapt, but I refused to lower my standards or abandon the principles I valued. I applied the discipline, resilience, and sense of duty I had cultivated in the military to my civilian endeavors. I also realized that while some civilians lacked the structured accountability found in the military, they were often eager to learn and grow when given the right guidance.

This inspired me to pour into others, both friends and colleagues, through mentorship. I volunteered for projects and helped others free of charge, not because I didn't value my time, but because I understood that money is temporary, while genuine guidance and support create lasting impact. Helping others find their potential became my mission, and the ripple effect of that effort has been incredibly rewarding.

The Challenges of Civilian Mentorship

Of course, there were challenges. Unlike the military, where quitting is rarely an option, civilian environments require contingency plans. I learned to always have a few replacements in my back pocket—people I could rely on when others faltered. While I hated the necessity of this, it reinforced my belief in the value of strong mentorship and leadership.

As a mentor, I focused on teaching resilience, discipline, and the importance of showing up—whether for others or for oneself. I emphasized the power of accountability, not just to a team but to one's own goals and values. My military experience became a toolset I could share, helping others build the confidence and determination they needed to succeed.

Why Mentorship Matters to Me

Mentorship is important to me because it's a way of paying it forward. The guidance I received in the Navy shaped the person I am today. It taught me how to lead, how to persevere, and how to bring out the best in others. By mentoring others, I am honoring that legacy and ensuring it continues.

Leadership, to me, is not about commanding others but about empowering them. It's about recognizing potential, fostering growth, and providing the tools and encouragement people need to reach their goals. Whether it's guiding a young sailor, helping a civilian friend navigate a new challenge, or volunteering in my community, I've seen firsthand how mentorship changes lives.

Tools for the Next Generation

As I continue to mentor and lead, I am committed to equipping the next generation with the tools they need to succeed. These include:

Resilience: Teaching others to push through challenges instead of quitting when the going gets tough.

Discipline: Emphasizing the importance of showing up on time and honoring commitments.

Service: Demonstrating that true leadership is about serving others, not oneself.

Long-term Vision: Helping others see that success is not about immediate rewards but about building a strong foundation for the future.

A Legacy of Leadership

The tools and values I brought from the Navy have served me well in civilian life. They have allowed me to inspire and guide others, to build lasting relationships, and to create a ripple effect of positive change. I am proud of the impact I have made and will continue to make as a mentor and leader.

Guiding the next generation is not just a responsibility; it is an honor. It is how we ensure that the values we hold dear—commitment, resilience, service, and integrity—live on. It is how we leave the world better than we found it. And for me, it is one of the most fulfilling parts of my journey.

Chapter 21: Fashion Shows and Milestones: A Platform for Creativity

Answering the Call for San Diego Design Week

When I was asked to close San Diego Design Week with a fashion show at the San Diego Made Factory, the honor felt deeply personal. It wasn't just another opportunity to showcase my designs; it was a moment to bring my journey full circle. As a Navy veteran, entrepreneur, and designer, this event allowed me to combine my past, present, and future into one unforgettable evening.

The significance of this show extended beyond fashion—it was a tribute to my 25 years of service in the United States Navy and a celebration of the transitions that have shaped my life. Every detail of the show was thoughtfully curated, down to the models who walked the runway. Each one of them played a unique role in my career, either during my time in the Navy or as part of my journey with Boss Lady Bling Blingy. These individuals were not just models; they were a living representation of the people who have inspired, supported, and stood by me throughout my path.

A Personal Journey on the Runway

The show began with me walking the runway in my Navy uniform—a powerful symbol of discipline, leadership, and resilience. It was a poignant reminder of the foundation upon which my journey was built. As the show progressed, each model brought a piece of that journey to life, wearing designs

that reflected specific moments in my career.

One model represented my early days of service, adorned in a bold yet elegant necklace that mirrored the strength and determination I needed to navigate the challenges of a military career. Another embodied the transition from service to civilian life, wearing a stunning crop top and accessories that celebrated reinvention and adaptability. Each piece of jewelry, every accessory, and each outfit told a part of my story, with the models giving life to the memories that shaped me.

The closing moment was a culmination of everything the show stood for. I walked out hand-in-hand with my dear friend and fellow designer, Oscar Romo, wearing one of my most cherished creations—my Blingy Dress White Blazer Jacket. This jacket, adorned with intricate beadwork and dazzling embellishments, symbolized my transition from the structured world of the Navy to the boundless creativity of fashion design. It was a deeply emotional moment, marking the end of my military service and the beginning of a new chapter as a designer and changemaker.

Supporting Fellow Designers and Expanding Horizons

While leading my own fashion shows has been a defining part of my journey, I have always found joy in collaborating with and supporting other designers. Behind the scenes, I've worked tirelessly to complement other collections with my handmade jewelry and accessories. From statement necklaces to dazzling chokers, my pieces have enhanced the visions of designers I deeply admire.

One of the highlights of this collaborative spirit was traveling to Lake Arrowhead for the GSFE Lady in Blue Fashion Show.

This prestigious event brought together women entrepreneurs and designers from across the region. Showcasing my jewelry in this beautiful setting was not just a professional milestone but also a celebration of the community of women who continue to inspire me.

The Significance of Milestones

Every milestone in my fashion journey reflects the interconnectedness of my life's experiences. Closing San Diego Design Week with my own show was a moment of immense pride, but it was also a deeply personal tribute. Dedicating the show to my 25 years of service gave it a profound meaning, allowing me to honor the journey that brought me to this moment.

Each design on that runway—from the first accessory to the final blazer jacket—was a piece of my story. Every model represented a chapter, a lesson, or a person who played a significant role in my career. These milestones are about more than personal achievement; they are about celebrating the relationships, struggles, and triumphs that shape us into who we are.

Walking Into the Future

The closing moments of the show will forever be etched in my memory: walking alongside Oscar Romo, wearing my Blingy Dress White Blazer Jacket, and seeing the models who represented the people and moments that defined my career. It was more than a fashion show—it was a celebration of community, creativity, and resilience.

As I move forward, I remain dedicated to using fashion as a platform for storytelling, empowerment, and collaboration. Each new show, every design, and every partnership strengthens my belief in the transformative power of art. With my bling, my creativity, and the unwavering support of my community, I am ready to embrace whatever comes next, knowing my journey has only just begun.

Chapter 22: Recognition and Awards

Boss Lady Bling Blingy Recognition and Awards: Designer of the Year and Beyond:

Awards and recognitions are more than accolades or shiny trophies. They are validations of a dream realized, a purpose fulfilled, and a mission amplified. For me, receiving titles like the Global ICONIC CHANGEMAKER of the 21st Century: Designer of the Year, the Ms. Elite Southern California International Woman of Achievement, the Kindness Award, and various recognitions from the State of California has not only elevated my brand, Boss Lady Bling Blingy, but also served as powerful affirmations of my life's mission: to inspire, empower, and uplift others through my art, service, and story.

A Journey to Recognition

From the moment I transitioned from military service into the world of design and entrepreneurship, my journey was never about fame or accolades. It was about transformation—turning challenges into opportunities, combat boots into blingy high heels, and military discipline into creative artistry. Yet, as the awards began to arrive, they became milestones marking my impact, motivating me to continue pushing boundaries and serving others.

Each recognition is a testament to the core values that guide my work: authenticity, empowerment, and a commitment to excellence. These awards are not just about me—they represent the incredible community I serve and the people whose lives I strive to touch through my designs, mentorship, and advocacy.

The Designer of the Year Award: A Defining Moment

Being honored as the Global ICONIC CHANGEMAKER of the 21st Century: Designer of the Year was a moment of profound validation. This award celebrated not just my artistic talent but also my dedication to creating unique designs that tell a story. My fashion pieces, from custom Navy retirement blazers to the dazzling handmade jewelry, are more than items—they are reflections of identity, resilience, and individuality.

This recognition reinforced my belief that design is a form of storytelling. Every client I serve, every piece I create, carries a narrative of empowerment and self-expression. It's this mission—helping others feel seen, valued, and celebrated—that fuels my passion and drives my business.

Ms. Elite Woman of Achievement Southern California 2024, Ms. Elite International Woman of Achievement 2025 Leadership in Action

Receiving the title of Ms. Elite Southern California International Woman of Achievement was a deeply humbling experience. This award represents more than my achievements as a designer; it acknowledges my role as a leader, mentor, and advocate for women and veterans.

As a retired Navy service member, I understand the challenges of transitioning to civilian life. This recognition reminded me of the importance of showing others—especially veterans and women—that their past experiences, no matter how difficult, can serve as steppingstones to greatness. It reinforced my commitment to empowering others to embrace their journeys and find their unique voices, just as I have through my brand.

State of California Recognitions: Impacting the Local Community

Being acknowledged by the State of California for my contributions to fashion, entrepreneurship, and the community was another profound honor. California has been both a backdrop and a springboard for my journey. It's where I've built my boutique, hosted fashion shows, and collaborated with local organizations to make a difference.

These awards are a reflection of the work I've done to uplift my community—whether it's mentoring aspiring designers, donating proceeds to Veteran Disability Services, or creating fashion pieces that celebrate diverse identities. They remind me that even as I dream big and aim globally, my impact begins right here at home.

The Kindness Award: A Legacy of Service

The Kindness Award is one of the recognitions closest to my heart. Kindness has always been a cornerstone of my mission, whether it's shown through the way I treat my clients, the causes I support, or the designs I create. This award affirmed that success isn't just about talent or hard work—it's about the heart and intention behind it all.

As someone who has navigated life's challenges, from serving in the military to building a business from the ground up, I understand the transformative power of kindness. It's this principle that guides every decision I make, every piece I design, and every interaction I have.

Awards as Catalysts for Growth

These accolades have done more than validate my work—they have catalyzed my growth as a designer, entrepreneur, and changemaker. They've opened doors to new opportunities, from collaborating with influential figures in the fashion world to hosting my own fashion shows, like the one that closed out San Diego Design Week.

They've also amplified my platform, allowing me to inspire more people through my story and mission. During the writing of this book, I realized these recognitions serve as reminders of the journey I've traveled and the lives I hope to touch through my words and designs.

Mission Reinforced: Empowerment Through Creativity

At the heart of all these awards is a simple yet powerful mission: to empower others through creativity. Each recognition serves as a reminder that my work is about more than fashion—it's about giving people the confidence to shine. Whether it's a veteran stepping into civilian life, a woman finding her unique voice, or a client celebrating a milestone, my designs are tools for transformation.

These accolades have reinforced my commitment to using my platform to inspire others. They remind me of the importance of sharing my story, supporting my community, and continuing to innovate in ways that uplift and empower.

Looking Ahead: Beyond the Awards

While I'm deeply grateful for the recognitions I've received, my journey doesn't end here. Awards are milestones, not destinations. They inspire me to dream bigger, work harder, and serve more people.

Looking ahead, I'm excited to continue expanding Boss Lady Bling Blingy—from launching new collections to growing my boutique's reach. I'm also committed to giving back, whether it's through mentorship, charitable donations, or creating designs that celebrate and uplift my clients.

The recognitions I've received are not just about celebrating past achievements—they are about building a legacy of impact, empowerment, and creativity.

A Note of Gratitude

None of these achievements would have been possible without the incredible support of my community. From clients who trust me with their visions to mentors who've guided me along the way, I am deeply grateful. I'd also like to thank organizations like the Global Society for Female Entrepreneurs for believing in my mission and recognizing my contributions.

To anyone reading this who has a dream or a mission, I want to say: believe in yourself, stay true to your values, and never stop striving for greatness. Every challenge you face is an opportunity to grow, and every success you achieve is a step toward making a difference.

Chapter 23: Rising from the Ashes

A Journey Through Divorce, Jail, Family, Business, and Self-Care

From 2022 to 2024, I lived through some of the hardest, most painful, and yet most transformative years of my life. I endured the deepest lows of disappointment, heartbreak, and self-doubt, while also uncovering moments of joy, gratitude, and resilience. If you've ever seen someone smile in public while privately holding themselves together by sheer will, you've seen me.

Many people see me as a woman who has it all. A retired military Senior Chief, a successful business owner, a mother, a model, a best-selling author, and an award-winning designer—on the outside, I represented strength, success, and stability. But what they didn't see was the storm raging within. I was broken, confused, and battling emotions that threatened to consume me.

Losing It All in the Blink of an Eye

When my husband of 22 years filed for divorce, I was devastated. The life I had built, the future I had envisioned, and the family I cherished felt like they were slipping away. It wasn't just the divorce that shook me—it was the timing. I was retiring from the military after 25 years of service, my son was graduating from college, and suddenly, I felt like everything I had built was crumbling all at once.

Family meant everything to me. I had worked hard to create a home and build a legacy. But during those moments, I didn't even recognize myself. The strong, determined woman I had always been was buried beneath layers of pain, exhaustion, and confusion. The pressure to maintain appearances was overwhelming.

On the outside, I was still the motivated Senior Chief who led by example. I still showed up as a loving mother, a thriving business owner, and a seemingly happy wife. But behind closed doors, I was dealing with depression, anger, and heartbreak.

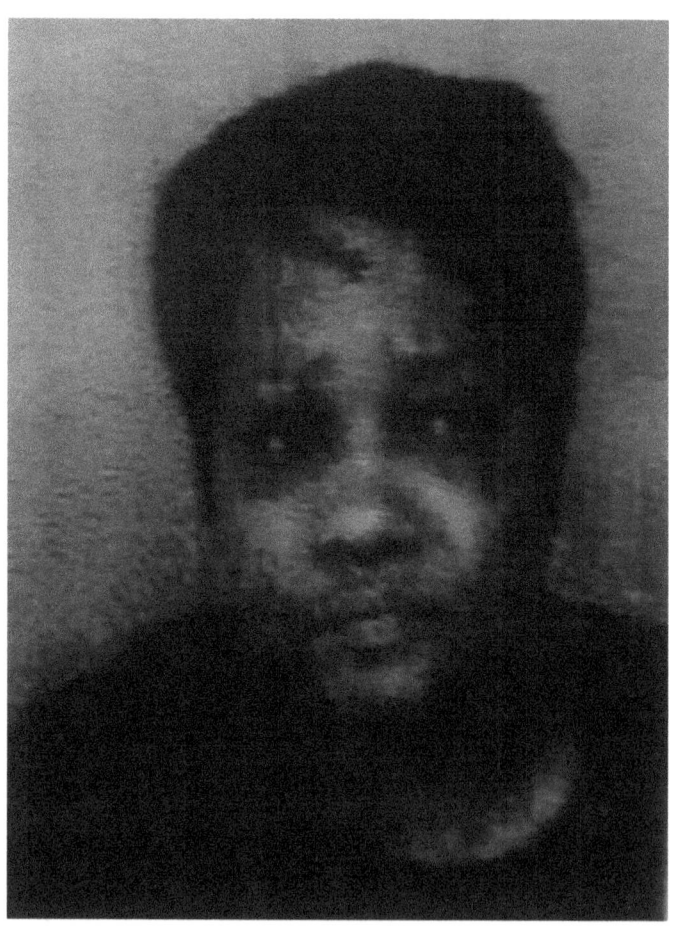

The Breaking Point

My breaking point came the day my husband pressed charges against me, and I found myself in jail. 11 hours and 37 minutes in a cold cell felt like an eternity. I kept asking myself, "How did I get here?" I couldn't believe the man I had spent nearly half my life with, the father of my son, would do something like this. But that day also became my turning point.

Every strong woman has a story—a moment when she hits rock bottom—and secrets that have shaped who she is. I've always been strong—as Navy Sailor, a businesswoman, a mother—but even the strongest among us fall. My fall came on an August morning at 4:50 a.m., in San Diego, California.

It's a memory so vivid that it feels like it happened yesterday. I was in the hotel lobby, sipping coffee and working on a customer's blinged-out order. The silence of the early morning was comforting, and I found solace in my craft. Then he walked in—my wasband.

He stumbled into the hotel, clearly drunk, not realizing I was downstairs. I watched him stumble to the elevator and then I waited about 30 minutes, trying to decide whether to head back upstairs. Finally, I packed up my things and made my way to our room, only to find myself locked out. He had latched the inside lock, leaving me stranded in the hallway.

It took me about 40 minutes to wake up my son so he could unlock the door. By the time I entered the room, I was tired, aggravated, and emotionally drained. I saw my wasband lying in my bed. Despite everything, I wanted to talk to him, to connect, to make sense of what we were going through. I called his name several times, but he didn't respond. Frustrated, I touched his feet, shaking him lightly to wake him up.

That's when everything spiraled out of control. He exploded with anger, yelling nonsensical things and throwing objects around the room. His aggression was overwhelming, and I felt trapped. I had reached my breaking point.

In a moment of desperation, I called the police for help, hoping they would de-escalate the situation. But when they arrived, he did the unthinkable—he pressed charges against me. He claimed I had assaulted him by touching his feet to wake him.

I couldn't believe it. I was in shock. How could this happen? I had never harmed him, never raised a hand against him. Yet, thanks to California's laws, his accusation was enough. I was handcuffed and taken away in front of my son.

Those 11 hours and 37 minutes in jail were the longest and most humiliating of my life. Sitting in that cell, I felt every emotion—anger, disrespect, sadness, shame, and, ultimately, clarity. This was the moment I knew my marriage was over. There was no going back, no fixing what had been broken for years.

Thankfully, California's District Attorney dropped the charges after it was determined that there had been no physical fight, and no one was harmed. It was a relief, but the damage was done. That moment became the catalyst for me to finally move forward with my divorce.

When I walked out of that jail, I wasn't the same woman who had walked in. I was determined to rebuild my life, to focus on myself and my son. I poured my energy into my business, Boss Lady Bling Blingy, and encouraged my son to finish his last semester of college. I let go of the weight of my broken marriage and chose to rise above it.

This chapter in my life taught me that even the strongest fall, but we don't stay down. We get up, brush ourselves off, and keep moving forward. It's okay to hit rock bottom because, from there, the only way is up.

Sitting in that cell, I realized I had been clinging to a marriage that no longer served me. I was fighting to hold onto a relationship that had eroded over time. I knew then that it was time to let go, not just of my marriage, but of the pain, disappointment, and anger I was carrying.

Walking out of that jail, I felt free—not because the situation was over, but because I had finally made peace with the fact that it was over.

Balancing Motherhood and Business Through the Storm

Even in my darkest moments, I had to keep going. My son was watching me, and I knew I had to set an example for him. Despite my personal struggles, I made it a priority to show up for him. His graduation was a proud moment and seeing him accomplish his dreams reminded me of my purpose.

At the same time, my business, Boss Lady Bling Blingy, demanded my attention. Running a business while going through a divorce was no small feat. Some days, I wanted to hide from the world, but I had clients, deadlines, and a brand to maintain. My boutique was more than just a job—it was my passion and my escape. Designing jewelry, styling clients, and creating custom pieces became my therapy.

I kept a straight face in public, even when I was crumbling inside. I poured my energy into my work, using it as a way to reclaim my identity and channel my emotions. My business became a symbol of my resilience.

The Power of Letting Go

Letting go of my marriage wasn't easy, but it was necessary. For years, I had stayed in the relationship because of the time we had invested. I told myself that being married for over 20 years was an accomplishment, even if I wasn't truly happy. But I realized that staying for the sake of appearances wasn't serving me or my son.

Letting go of the hurt and focusing on myself allowed me to heal. I stopped settling for less than I deserved, and I rediscovered my worth. I began to embrace life on my terms, and for the first time in years, I felt free.

Self-Care as a Survival Tool

One of the most important lessons I learned during this time was the power of self-care. As women, especially Black women, we're often expected to be strong and resilient no matter what we're going through. But that strength can come at a cost if we don't take the time to nurture ourselves.

Self-care became my lifeline. I prioritized my physical, mental, and emotional health in ways I never had before. I started with small steps—going for walks, journaling, and taking time to reflect. I also leaned on therapy and my faith, which gave me the strength to move forward.

I focused on my appearance, not for vanity, but as a way to reclaim my confidence. I started dressing up more, experimenting with new styles, and embracing my femininity. I reminded myself every day that I was beautiful, worthy, and deserving of love—not just from others, but from myself.

Rediscovering Happiness and Gratitude

As I began to heal, I found happiness in unexpected places. My relationship with my son grew stronger as we navigated this new chapter together. I found joy in my business, creating pieces that brought smiles to my clients' faces. I embraced the freedom that came with letting go of a toxic disrespectful relationship.

I also found gratitude in the lessons I had learned. The challenges I faced made me stronger and more self-aware. They taught me the importance of boundaries, self-worth, and the power of resilience.

Inspiring Others Through My Journey

My story isn't just about surviving divorce or overcoming adversity—it's about thriving in the face of it. I've learned that life's challenges don't define us; how we respond to them does.

As a mother, business owner, and Black woman, I've faced moments that tested me to my core. But I came out on the other side stronger, wiser, and more determined than ever. I want other women to know if you're going through something similar or have been down this road, you too, can rise from the ashes.

Final Thoughts

Going through divorce, retiring, balancing motherhood, running a business, and prioritizing self-care taught me that we are capable of so much more than we realize. It's okay to feel broken, to take time to grieve, and to ask for help. But it's also important to remember that healing and happiness are possible.

Looking back, I can say that 2022 to 2024 were some of the hardest difficult years of my life, but they were also the most transformative. I learned to let go of what no longer served me, embrace my worth, and step into the next chapter of my life with confidence and gratitude.

To any woman reading this, know that you are not alone. Whether you're facing divorce, balancing family, and career, or trying to find yourself again, you are capable of rising above it all. You are worthy, you are strong, and you are enough but most of all you're Blingy BEAUTIFUL!

This chapter of my life was a test, but it also became my testimony. And for that, I am forever grateful.

Chapter 24: The Legacy of Boss Lady Bling Blingy: Inspiring Strength, Style, and Resilience

When I started Boss Lady Bling Blingy, it wasn't just about creating beautiful, custom jewelry, accessories, and fashion pieces; it was about building a legacy. For me, the brand has always been more than a business—it is a testament to resilience, creativity, and empowerment. It's a tribute to where I've come from and a beacon of hope for others, particularly veterans and women, to follow their own paths of self-discovery, purpose, and success.

The Seed of Inspiration

The journey to becoming the Boss Lady wasn't linear. My 25 years of service in the U.S. Navy taught me discipline, leadership, and the importance of working under pressure, but it also planted seeds of self-doubt and weariness. When I decided to transition out of the military, I realized that I needed to rediscover who I was outside of the uniform. What could I offer the world? Who was Latasha Fennell beyond the titles I had carried?

That journey of rediscovery became the foundation for Boss Lady Bling Blingy. It taught me that life after the military—or any significant chapter—requires reinvention. You aren't leaving something behind; you're building on the foundation of everything you've learned, endured, and overcome. I want my brand to inspire others to see their transitions in life as opportunities to shine brighter than ever.

A Legacy of Empowerment

At the core of Boss Lady Bling Blingy is the belief that we all possess a unique light waiting to be unleashed. My designs—whether it's a bold Blingy Blazer Jacket or a delicate vintage-inspired necklace—are not just about aesthetics. They are about creating pieces that allow people to feel their most confident and authentic selves. When someone wears something, I've created, my hope is that they feel seen, celebrated, and empowered.

For veterans, this empowerment is especially critical. The transition from military service to civilian life can feel overwhelming and isolating. Many of us struggle with finding our place in a world that doesn't operate like the one we've known. Through my work, I want veterans to know that they are not alone in this journey. Their skills, discipline, and experiences are valuable, and their contributions to society can extend far beyond their military service.

For women, my legacy is rooted in the belief that we are powerful beyond measure. As women, we often wear many hats—mother, wife, leader, friend—and it's easy to lose sight of our own dreams in the process of serving others. Boss Lady Bling Blingy is here to remind women to prioritize themselves, embrace their creativity, and never be afraid to stand out.

Inspiration Through Authenticity

One of the most fulfilling parts of this journey has been sharing my story. From my beginnings in the military to my leap into entrepreneurship, I've learned that authenticity is what truly inspires others. The moments when I've been most vulnerable—whether stepping onto a runway in my Navy uniform or designing a piece that reflects my own struggles and triumphs—are the moments that resonate most with people.

I want my brand to inspire others to be unapologetically themselves. Whether you're transitioning out of a career, launching a business, or simply trying to find your footing in life, know that your story has power. The more we lean into our authenticity, the more we open doors for others to do the same.

Building Confidence, One Design at a Time

Confidence is a recurring theme in everything I create. When I design a Diamond Push-up Bustier or a custom Navy Retirement Blazer, I'm not just thinking about how it looks— I'm thinking about how it makes the person wearing it feel. Clothing and accessories have the power to transform how we see ourselves. They can reflect our inner strength and be a catalyst for the confidence we need to take on the world.

This is especially important for veterans and women, who often face societal pressures that make them doubt their worth or abilities. For veterans, it might be the stereotype that life after service is a downward spiral. For women, it might be the expectation to downplay their ambitions or shrink themselves to fit into a mold. Through my brand, I want to shatter these narratives and remind people that they are capable of greatness.

The Ripple Effect of Giving Back

A significant part of my legacy is tied to giving back. Supporting veterans through my work with organizations like VDS (Veteran Disability Services) is one of the ways I honor those who have served. When I see a veteran wear one of my designs with pride, it's not just about the clothing—it's about the confidence and hope it instills. It's about reminding them that they have a community that values and supports them.

For women, my hope is that the legacy of Boss Lady Bling Blingy inspires them to uplift others as they rise. Whether it's through mentorship, collaboration, or simply being a source of encouragement, we all have the power to create ripples of positive change. My journey has been enriched by the women who have supported and believed in me, and I want to pay that forward through everything I do.

Inspiring Creativity and Courage

Another cornerstone of my legacy is creativity. I often tell people that Boss Lady Bling Blingy wasn't born out of perfection—it was born out of passion and courage. I had to be willing to take risks, make mistakes, and learn as I went. This is the message I want to pass on: you don't have to have everything figured out to start. Whether it's designing your first piece of jewelry or launching a business, what matters most is that you take that first step.

Creativity requires courage. It's about trusting your vision and not being afraid to stand out. I want veterans and women to know that their creativity is a gift, and the world needs what they have to offer. Every design, every idea, and every dream have the potential to inspire and uplift others.

From Combat Boots to Blingy High Heels

From Combat Boots to Blingy High Heels, encapsulates the journey of transformation and resilience that defines my story and brand. It's a journey I hope will inspire others to embrace their own transitions with grace and determination. Whether you're a veteran stepping into civilian life or a woman reclaiming her dreams, know that your path is unique and filled with endless possibilities.

Boss Lady Bling Blingy is my way of showing the world that we can be strong and stylish, resilient, and radiant. It's a reminder that life is not about leaving behind who you were but about carrying the lessons of your past into the future with pride.

A Call to Action

To every veteran reading this: your service is a foundation, not a limitation. You have the tools, discipline, and resilience to create a fulfilling and meaningful life beyond the military. Your story matters, and the world needs your contributions.

To every woman reading this: your dreams are valid, and your voice deserves to be heard. Don't be afraid to take up space, to shine, and to pursue the life you envision for yourself. You are more powerful than you realize, and your creativity and courage can move mountains.

The legacy of Boss Lady Bling Blingy is one of empowerment, creativity, and resilience. It's a legacy I hope will inspire others to dream bigger, rise higher, and shine brighter. Remember: the journey isn't about perfection—it's about passion, persistence, and believing in the light within you. Shine on.

Chapter 25: Walking in Your Own High Heels

Reflections

As I pen this concluding chapter, I am overcome with the weight and wonder of my journey—from the structure of combat boots to the confidence of blingy high heels. This is more than a transition in footwear; it's a profound transformation of identity, perspective, and purpose. Every step, stumble, and stride has shaped me into the woman I am today—a woman rooted in authenticity, empowered by transformation, and fueled by purpose.

Embracing Transformation

Transformation is rarely a straightforward journey. It's a path woven with moments of triumph and heartache, clarity, and confusion. When I hung up my combat boots, I wasn't just stepping away from a military career; I was stepping into the unknown—a space where I had to rediscover myself beyond the uniform. Those boots represented discipline, strength, and service, but they also carried the weight of expectations that sometimes stifled my true essence.

Stepping into my high heels symbolized a bold reclamation of my identity. It wasn't about abandoning who I was but about embracing who I could become. Transformation is often uncomfortable because it demands growth. It asks us to let go of the familiar and trust in the possibilities of the unknown. For me, that meant navigating the unpredictable worlds of fashion

and entrepreneurship—a stark contrast to the regimented life I once knew.

Through this journey, I learned that transformation is not about erasing the past but integrating it. The resilience, discipline, and integrity I cultivated in the Navy are now the bedrock of my creative pursuits and entrepreneurial ventures. Each design I create and every milestone I achieve reflects the lessons of perseverance and adaptability that my years of service instilled in me.

Living Authentically

Authenticity is the cornerstone of a fulfilling life. For many years, I wore the roles assigned to me—a service member, a leader, a caregiver. But as I leaned into my passion for creativity and self-expression, I realized that true authenticity means honoring every facet of who you are, even those parts that feel unconventional or misunderstood.

Walking in my own high heels taught me that authenticity isn't a fixed destination; it's a dynamic process of self-discovery. It's about aligning your actions, values, and dreams with your innermost truths. For me, this means designing bold, statement-making pieces that reflect my journey and encourage others to embrace their uniqueness. It means standing unapologetically in my story and using it to inspire others to do the same.

Authenticity has also deepened my purpose. Through my work, I strive to empower others, particularly veterans, to redefine their lives after service. It's not just about the jewelry I design or the blazers I craft; it's about creating pieces that resonate with confidence, resilience, and individuality.

Living with Purpose

Purpose is the compass that guides us through life's uncertainties. It's the thread that ties our experiences together, giving meaning to our struggles and direction to our dreams. My purpose has always been rooted in service—first to my country and now to my community.

Through Boss Lady Bling Blingy, I've found a platform to uplift, empower, and give back. Whether it's mentoring fellow veterans, creating custom designs that celebrate life's milestones, or donating to causes close to my heart, my work is a reflection of the purpose that drives me.

Purpose is what keeps me grounded during challenging times. It reminds me that my journey is bigger than me—it's about creating a legacy of transformation and empowerment. It's about showing others that they, too, can walk confidently in their own high heels, whatever that may look like.

Your Journey

As you close this book, I invite you to reflect on your own journey. What have you overcome? How have you grown? And most importantly, what is the purpose that fuels your path?

Walking in your own high heels isn't about emulating someone else's journey—it's about embracing your own. Whether your steps are in combat boots, sneakers, or stilettos, they are uniquely yours. Honor them. Celebrate them. And trust that every step is leading you closer to the person you're meant to be.

Final Reflections

As I look back on my journey, I am filled with gratitude. Every challenge I faced and every triumph I celebrated was a steppingstone to this moment of clarity, purpose, and authenticity. The road hasn't always been easy, but it has been worth every step.

To you, the reader, I offer this: Transformation is always possible. Authenticity is always within reach. Purpose is the gift you give yourself and the world. Step boldly into your journey, live unapologetically, and trust in the power of your story.

This chapter may be closing, but the story is far from over. This is the beginning of your next chapter—a chapter filled with courage, joy, and the unshakable belief in your power to transform, inspire, and thrive.

Keep walking, keep shining, and never stop believing in the extraordinary journey that lies ahead.

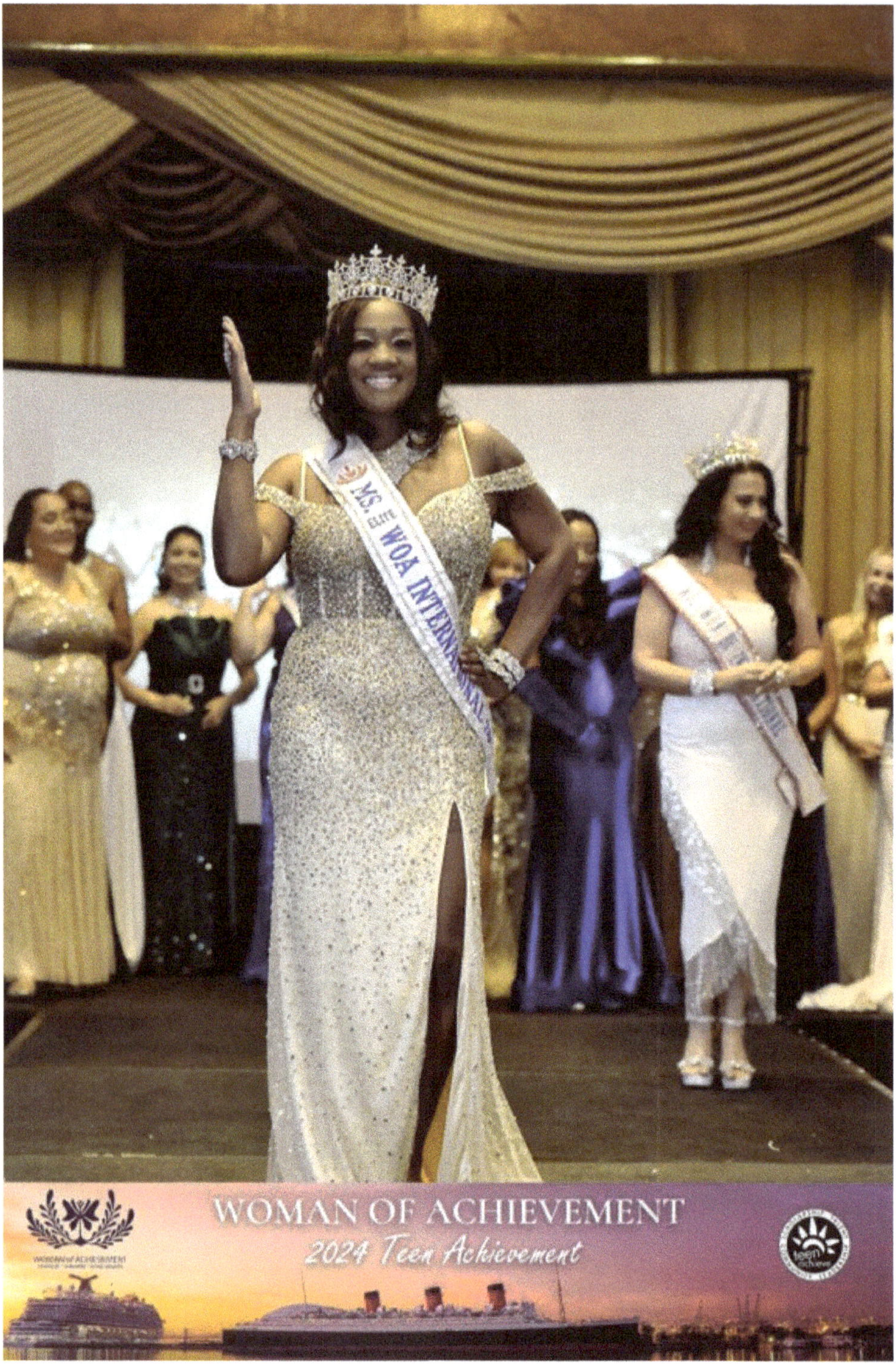
WOMAN OF ACHIEVEMENT
2024 Teen Achievement

Amb. Dr. LATASHA FENNELL (h.c.)

About The Author

"IT WAS NEVER ON ME ...
BUT HAS ALWAYS BEEN IN ME!

Latasha Fennell is a trailblazing leader whose life journey embodies resilience, empowerment, and service. A retired Senior Chief Petty Officer with 25 years of honorable service in the United States Navy, Latasha exemplified integrity, discipline, and unwavering commitment throughout her military career. Her tenure in the Navy not only prepared her as a leader but also equipped her with the courage and determination to face life's challenges head-on.

Following her retirement, Latasha embraced entrepreneurship, founding Boss Lady Bling Blingy Boutique, a thriving business renowned for its custom handmade jewelry, accessories, and bespoke fashion. From her signature Blingy Blazer Jackets to unique vintage-style necklaces, Latasha's designs celebrate individuality, creativity, and confidence. Her boutique is more than a business—it is a movement inspiring woman to own their style and step boldly into their power.

Latasha's outstanding contributions have earned her numerous accolades, including the President's Volunteer Service Award (Gold), an honorary doctorate in Humanitarianism, the Global ICONIC Changemaker Award 2024, and several prestigious honors from the state of California. She is also the proud recipient of the esteemed title Ms. Elite Southern California

2024, Ms. Elite International Woman of Achievement 2025, recognizing her leadership, philanthropy, and dedication to uplifting others.

BOSS LADY BLING BLINGY

SAN DIEGO BUSINESS JOURNAL

BLACK LEADERS OF INFLUENCE **HONOREE** 2025

LATASHA FENNELL

CEO of Boss Lady Bling Blingy Boutique

CONGRATULATIONS
to **Boss Lady Bling Blingy Boutique CEO Latasha Fennell**
for being a 2025 Black Leader of Influence.
We're all so proud of you!

bossladyblingblingy.com

Latasha Fennell is one of the Black leaders Honoree of influence in San Diego in 2025, as named by the San Diego Business Journal. Latasha is the owner and creator of Boss Lady Bling Blingy Boutique.

The San Diego Business Journal's Black Leaders of Influence award recognizes Black leaders in San Diego who have made significant contributions to the city's business community.

She is a passionate advocate for veterans. Latasha is deeply committed to helping service members transition successfully into civilian life. Through her boutique, she donates a portion of proceeds to Veteran Disability Services (VDS), ensuring continued support for those who served.

As a best-selling author, motivational speaker, and mentor, Latasha captivates audiences with her transformative story. Her upcoming book, "From Combat Boots to Blingy High Heels," chronicles her evolution from military life to entrepreneurial success, offering readers a roadmap to resilience and empowerment.

Above all, Latasha is a proud mother to her handsome son Trenton Fennell, drawing strength and inspiration from her role as a parent. She continues to inspire countless individuals with her unwavering commitment to service, creativity, and empowerment. Latasha Fennell is a changemaker whose life's work exemplifies the courage to embrace new beginnings and the power of living with purpose and style.

BOSS LADY
BLING BLINGY

www.ingramcontent.com/pod-product-compliance
Lightning Source LLC
Chambersburg PA
CBHW051520120626
46551CB00012B/1002

* 9 7 9 8 9 9 2 7 5 2 5 0 2 *